Amazon
Fire Phone

the missing manual®
The book that should have been in the box

Preston Gralla

O'REILLY®

Beijing | Cambridge | Farnham | Köln | Sebastopol | Tokyo

Amazon Fire Phone: The Missing Manual
by Preston Gralla

Copyright © 2015 Preston Gralla. All rights reserved.
Printed in the United States of America.

Published by O'Reilly Media, Inc., 1005 Gravenstein Highway North, Sebastopol, CA 95472.

O'Reilly books may be purchased for educational, business, or sales promotional use. Online editions are also available for most titles (*https://www.safaribooksonline.com*). For more information, contact our corporate/institutional sales department: 800.998.9938 or *corporate@oreilly.com*.

December 2014:   First Edition.

Revision History for the First Edition:

2014-12-08   First release

See *http://www.oreilly.com/catalog/errata.csp?isbn=0636920034544* for release details.

ISBN: 978-1-491-91123-5
[LSI]

Contents

PART IV **Appendixes**

The Missing Credits

About the Author

Preston Gralla is the author of more than 40 books that
have been translated into 20 languages, including *Galaxy
S5: The Missing Manual*, *Windows 8 Hacks*, *NOOK HD:
The Missing Manual*, *Galaxy Tab: The Missing Manual*, *The
Big Book of Windows Hacks*, *How the Internet Works*,
and *How Wireless Works*. He is a contributing editor to
Computerworld, a blogger for ITWorld, and was a founding editor and then editorial director of *PC/Computing*,
executive editor for CNET/ZDNet, and the founding
managing editor of *PC Week*.

He has written about technology for many national newspapers and magazines, including *USA Today*, the *Los Angeles Times*, *The Dallas Morning News*
(for which he wrote a technology column), *PC World*, and numerous others. As
a widely recognized technology expert, he has made many television and radio
appearances, including on the CBS *Early Show*, MSNBC, ABC *World News Now*,
and National Public Radio. Under his editorship, *PC/Computing* was a finalist
for General Excellence in the National Magazine Awards. He has also won the
"Best Feature in a Computing Publication" award from the Computer Press
Association.

Gralla is also the recipient of a Fiction Fellowship from the Massachusetts
Cultural Council. He lives in Cambridge, Massachusetts, with his wife (his two
children have flown the coop). He welcomes feedback about his books by email
at *preston@gralla.com*.

About the Creative Team

Nan Barber (editor) has worked with the Missing Manual series since its inception—long enough to remember booting up her computer from a floppy disk. Email: *nbarber@oreilly.com*.

Kara Ebrahim (production editor) lives, works, and plays in Cambridge, Mass. She loves graphic design and all things outdoors. Email: *kebrahim@oreilly.com*.

Brian Sawyer (technical reviewer) is a Senior Editor at O'Reilly Media, where he manages the Missing Manuals division. He is also the author of *Fire Phone: Out of the Box* and *Kindle Fire: Out of the Box* and coauthor of *NOOK Tablet: Out of the Box* and *Best Android Apps*. Email: *bsawyer@oreilly.com*.

Julie Van Keuren (proofreader) quit her newspaper job in 2006 to move to Montana and live the freelancing dream. She and her husband, M.H. (who is living the novel-writing dream), have two sons, Dexter and Michael. Email: *little_media@yahoo.com*.

Ron Strauss (indexer) specializes in the indexing of information technology publications of all kinds. Ron is also an accomplished classical violist and lives in Northern California with his wife and fellow indexer, Annie, and his miniature pinscher, Kanga. Email: *rstrauss@mchsi.com*.

Acknowledgements

Many thanks go to my editor, Nan Barber, who not only patiently shepherded this book through the lengthy writing and publishing process, but provided valuable feedback and sharpened my prose. Thanks also go to Brian Sawyer, for thinking of me for this book.

I'd also like to thank all the other folks at O'Reilly who worked on this book, especially Kara Ebrahim, Rebecca Demarest, Julie Van Keuren, and Ron Strauss.

—Preston Gralla

The Missing Manual Series

Missing Manuals are witty, superbly written guides to computer products that don't come with printed manuals (which is just about all of them). Each book features a handcrafted index and cross-references to specific pages (not just chapters).

Recent and upcoming titles:

Access 2013: The Missing Manual by Matthew MacDonald

Adobe Edge Animate: The Missing Manual by Chris Grover

Buying a Home: The Missing Manual by Nancy Conner

Creating a Website: The Missing Manual, Third Edition by Matthew MacDonald

CSS3: The Missing Manual, Third Edition by David Sawyer McFarland

Dreamweaver CS6: The Missing Manual by David Sawyer McFarland

Dreamweaver CC: The Missing Manual, Second Edition by David Sawyer McFarland and Chris Grover

Excel 2013: The Missing Manual by Matthew MacDonald

FileMaker Pro 13: The Missing Manual by Susan Prosser and Stuart Gripman

Flash CS6: The Missing Manual by Chris Grover

Galaxy Tab: The Missing Manual by Preston Gralla

Galaxy S5: The Missing Manual by Preston Gralla

Google+: The Missing Manual by Kevin Purdy

HTML5: The Missing Manual, Second Edition by Matthew MacDonald

iMovie: The Missing Manual by David Pogue and Aaron Miller

iPad: The Missing Manual, Seventh Edition by J.D. Biersdorfer

iPhone: The Missing Manual, Eighth Edition by David Pogue

iPhone App Development: The Missing Manual by Craig Hockenberry

iPhoto: The Missing Manual by David Pogue and Lesa Snider

iPod: The Missing Manual, Eleventh Edition by J.D. Biersdorfer and David Pogue

iWork: The Missing Manual by Jessica Thornsby and Josh Clark

JavaScript & jQuery: The Missing Manual, Third Edition by David Sawyer McFarland

Kindle Fire HD: The Missing Manual by Peter Meyers

Living Green: The Missing Manual by Nancy Conner

Microsoft Project 2013: The Missing Manual by Bonnie Biafore

Motorola Xoom: The Missing Manual by Preston Gralla

NOOK HD: The Missing Manual by Preston Gralla

Office 2011 for Macintosh: The Missing Manual by Chris Grover

Office 2013: The Missing Manual by Nancy Conner and Matthew MacDonald

OS X Mavericks: The Missing Manual by David Pogue

OS X Yosemite: The Missing Manual by David Pogue

Personal Investing: The Missing Manual by Bonnie Biafore

Photoshop CS6: The Missing Manual by Lesa Snider

Photoshop CC: The Missing Manual, Second Edition by Lesa Snider

Photoshop Elements 13: The Missing Manual by Barbara Brundage

PHP & MySQL: The Missing Manual, Second Edition by Brett McLaughlin

QuickBooks 2015: The Missing Manual by Bonnie Biafore

Switching to the Mac: The Missing Manual, Mavericks Edition by David Pogue

Windows 7: The Missing Manual by David Pogue

Windows 8: The Missing Manual by David Pogue

WordPress: The Missing Manual, Second Edition by Matthew MacDonald

Your Body: The Missing Manual by Matthew MacDonald

Your Brain: The Missing Manual by Matthew MacDonald

Your Money: The Missing Manual by J.D. Roth

For a full list of all Missing Manuals in print, go to *www.missingmanuals.com/ library.html.*

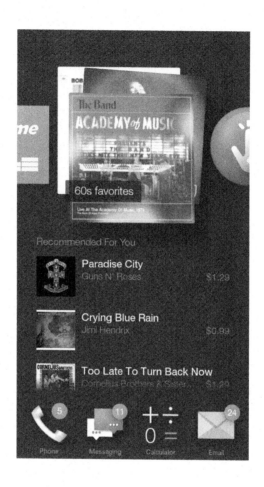

Introduction

THE FIRE PHONE WAS four years in the making. An early prototype was shown to AT&T in 2011, and it went on sale exclusively at AT&T in late July 2014. At first glance, and when you hold it in your hand, it doesn't seem different from any other smartphone you might buy—essentially a plain, black slab. However, once you get to the Lock screen, you immediately see a difference, because the phone's Dynamic Perspective lets you look at the image in full 3D. You can even change the perspective of your view simply by moving your hand.

Like most smartphones, the Fire phone gives you high-speed Internet access, runs games and apps, lets you take high-resolution photos and HD videos, gives you immediate access to your favorite social networks, handles any email you can throw at it, lets you watch free TV and movies and listen to free music, and keeps you in touch by phone and text. As you'll also learn, the Fire phone connects you directly to Amazon's vast marketplace. Thanks to the new Firefly technology, all you have to do is point the phone at something, and you'll be able to buy the product—or book, or movie, or TV show—from Amazon with a tap. The phone also has a unique help feature called Mayday that lets you get tech support live via video from Amazon. And when you buy the phone, you also get a one-year free subscription to Amazon Prime.

This book will help you get the most out of your Fire, and there's a lot you can get out of it. Whether you're just looking to get started or want to dig deep into the phone's capabilities, this book has you covered.

NOTE When you first use the Fire, you can take a Quick Start tour to get acquainted with all the phone's features. It's well worth it, so give it a try.

About the Amazon Fire Phone

AMAZON HAS STARTED WITH a solid piece of hardware and layered some innovative software on top of it. The Fire phone has a 4.7-inch, high-resolution screen; a 13-megapixel camera for high-res photos and video; and a front-facing 2.1-megapixel camera for selfie-taking. Its brain is a fast 2.2 GHz four-core processor. For keeping you connected, the Fire has antennas for Bluetooth, WiFi, and GPS. It has access to the speedy 3G and 4G networks, which let you talk, text, and surf the Web almost anywhere in the U.S.

> **NOTE** This book was written based on the AT&T version of the Amazon Fire phone. The Fire was initially released only for AT&T. Versions from other carriers—if they become available—may have minor variations in what you see onscreen.

Amazon took Google's Android operating system and did some very serious tweaking to it, so that most of the time it bears only a passing resemblance to Android. (The operating system, the Fire OS, is similar to the one used by other Amazon devices, including the Kindle Fire line of tablets, as discussed in the box on page 20.)

Put it all together, and you can do just about anything. You can get turn-by-turn directions, check weather and traffic, and identify landmarks. You can manage your email and calendar. You can take pictures and share them on Facebook. You can turn the Fire into a WiFi hotspot for getting other devices online. And you can get content from Amazon's vast ecosystem of movies, TV shows, books, and music.

Oh, and it's also a darn good phone with great sound quality and all the calling features you could ask for.

You could figure out how to make the most of all these features on your own, but by that time there'd be a whole new generation of smartphones to learn. This book will put you on the fast track to all the Fire's magic.

Buying and Contract Options

To buy a Fire phone, point your browser to *www.amazon.com/firephone*. You'll land on a page that describes all the phone's features and buying options. You may also find the phone in stock at your local AT&T store or a retailer like Best Buy.

As this book went to press, the Fire phone was available only from AT&T, and it had a variety of pricing plans. If you sign up for a two-year contract, you get the Fire for 99¢ (that's right, 99 *cents*). If you buy it with no contract, it will cost you

$449. (Note that even if you buy it with no contract, you can still use the phone only on the AT&T network.)

If you sign up for AT&T's Next contract, things get more complicated. You'll pay nothing upfront for the phone, but you'll have to pay something every month for it—it's like buying on the installment plan. Depending on which Next contract you choose, you'll pay either $22.50 a month for 20 months, or $18.75 a month for 24 months.

What's Unique about the Fire Phone

WHAT MAKES THE FIRE special is a combination of hardware, software, and Amazon's vast ecosystem:

- **Amazon Prime.** Buy the Fire phone, and you get a free year of Amazon Prime, which normally costs $99. That gets you free shipping on most of what you buy on Amazon, plus free movies, TV shows, and music.

- **Dynamic Perspective.** This feature is unique among smartphones. You can control the phone, take actions, and even play games by merely moving your eyes or hand.

- **Firefly.** Firefly is a feature that looks at the world and the objects in it, identifying things you can buy—text, posters, artwork, phone numbers, email addresses, and more. And it does more than look. It can listen as well and recognize music, movies, and TV shows by what it hears. Naturally, most of what you find you can buy on Amazon.

- **Mayday.** Confused about some aspect of the Fire phone? Want some help using Amazon's services? Press the Mayday button and you get live help, right on the phone.

- **Amazon Cloud.** The Fire phone also gives you an on-ramp to the Amazon cloud. There you'll find any content that you've bought on another Amazon device, and you can then stream it to your Fire phone, or download it. You'll also find photos and videos from other devices, music that you've uploaded to the cloud, and more.

> **NOTE** Amazon giveth, and Amazon taketh away. Here's something you don't get on the Fire phone: Google apps. The Gmail app, Google Maps, Google Plus, and so on, are not available for the Fire phone. (You can, however, use Gmail with the Fire's built-in Email app; see page 193.)

About This Book

THERE'S AN ENTIRE WORLD to explore in the Fire phone, and the little leaflet that comes in the box doesn't begin to give you all the help, advice, and guidance you need. So this book is the manual that should have accompanied the Fire phone.

There's a chance that since this book was written, there have been some changes to the Fire phone. To help keep up to date about them, head to this book's Errata/Changes page at *http://oreil.ly/1zyib36*.

About the Outline

FIRE PHONE: THE MISSING Manual is divided into four parts, each of which has several chapters:

- **Part I: The Basics.** Provides a guided tour of the Fire, from the power button to the unique features like Dynamic Perspective, Firefly, Mayday, and Amazon Prime. You'll learn all about making phone calls and text messages.

- **Part II: Media.** Gives you the rundown on using the Fire for reading books and magazines, taking pictures, recording and watching video, and playing and managing your music. And, because the Fire phone can connect to a computer like any USB device, you'll learn how to transfer your media files and other documents between the phone and computer.

- **Part III: The Fire Phone Online.** Tells you everything you need to know about the Fire phone's online talents. You'll find out how to get online, either over your service provider's network or a WiFi hotspot; see how you can turn your Fire into a portable WiFi hotspot; master email; browse the Web; and sync your Fire phone with Facebook and Twitter. You'll also learn about Amazon's mapping app, how to navigate using GPS, and how to find any location in the world. And if you still want more, you'll learn how to download and use the wide variety of apps from the Amazon Appstore.

- **Part IV: Appendixes.** Here are your Fire phone reference resources. Appendix A shows you how to activate your Fire. Appendix B shows what kind of accessories you can get for your Fire phone, such as cases, chargers, and screen protectors. Appendix C offers plenty of help troubleshooting issues with the phone's operation, and Appendix D covers all the Fire phone's settings.

About→These→Arrows

IN THIS BOOK AND in the entire Missing Manual series, you'll find instructions like this one: Tap Settings→Call Settings→"Voicemail settings." That's a short-hand way of giving longer instructions like this: "Tap the Settings button. From the screen that opens, tap Call Settings. And from the screen that opens after that, tap 'Voicemail settings.'"

It's also used to simplify instructions you'll need to follow on your PC or Mac, like File→Print.

About the Online Resources

AS THE OWNER OF a Missing Manual, you've got more than just a book to read. Online, you'll find example files so you can get some hands-on experience, as well as tips, articles, and maybe even a video or two. You can also communicate with the Missing Manual team and tell us what you love (or hate) about the book. Head over to *www.missingmanuals.com*, or go directly to one of the following sections.

Missing CD

So you don't wear down your fingers typing long web addresses, the Missing CD page offers a list of clickable links to the websites mentioned in this book. Go to *www.missingmanuals.com/cds/fptmm* to see them all neatly listed in one place.

Registration

If you register this book at *www.oreilly.com*, you'll be eligible for special offers—like discounts on future editions of *Fire Phone: The Missing Manual*. Registering takes only a few clicks. To get started, type *http://oreilly.com/register* into your browser to hop directly to the Registration page.

Feedback

Got questions? Need more information? Fancy yourself a book reviewer? On our Feedback page, you can get expert answers to questions that come to you while reading, share your thoughts on this Missing Manual, and find groups for folks who share your interest in the Fire phone. To have your say, go to *www.missing-manuals.com/feedback*.

Errata

In an effort to keep this book as up to date and accurate as possible, each time we print more copies, we'll make any confirmed corrections you've suggested. We also note such changes on the book's website, so you can mark important corrections into your own copy of the book, if you like. Go to *http://oreil.ly/1zyib36* to report an error and to view existing corrections.

Safari® Books Online

SAFARI® BOOKS ONLINE IS an on-demand digital library that lets you easily search over 7,500 technology and creative reference books and videos to find the answers you need quickly.

With a subscription, you can read any page and watch any video from our library online. Read books on your cellphone and mobile devices. Access new titles before they're available for print, and get exclusive access to manuscripts in development and post feedback for the authors. Copy and paste code samples, organize your favorites, download chapters, bookmark key sections, create notes, print out pages, and benefit from tons of other time-saving features.

O'Reilly Media has uploaded this book to the Safari Books Online service. To have full digital access to this book and others on similar topics from O'Reilly and other publishers, sign up for free at *https://www.safaribooksonline.com*.

The Basics

You'll learn to:

- Operate the Fire phone's hardware

- Navigate the Carousel, Apps Grid, and side panels

- Use Dynamic Perspective

- Control the Fire with gestures—including one-handed gestures

CHAPTER 1

The Guided Tour and Special Features

THE FIRST TIME YOU hold the Fire phone, you'll want to put it through its paces: admiring its unique 3D Lock screen, trying out Firefly (which identifies objects and gives you more information about them), and doing all kinds of amazing things by merely tilting the phone. Oh, yes, and you'll want to also check your email, run apps, read books, and browse Amazon. Before you do that, though, you need a solid understanding of how the Fire phone works and familiarity with all its different parts. You'll want to know where all its buttons, keys, and ports are located, for example—not to mention how to get to your Home screen.

Power/Lock Button

AT THE FIRE PHONE'S top left, you'll find a small, rectangular black button. It's a hardworking button with multiple functions. Press it, and it opens your Lock screen. Press and release it when the phone is turned on and active, and it puts the phone into Standby mode. When you press *and hold* the button when the phone is turned on, a screen appears that lets you power off or restart the phone.

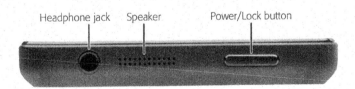

Headphone jack   Speaker   Power/Lock button

Locking the Screen

When you put the Fire phone on Standby using the Power/Lock button, the screen stops responding to touch. It blacks out, indicating that the screen is *locked*. Always lock the screen before putting the Fire phone in your pocket or bag to avoid accidental screen taps and embarrassing unintended phone calls. In fact, every time you leave the phone untouched for a certain amount of time—as little as 30 seconds to as much as 30 minutes (which you can adjust, as described on page 317)—the screen automatically locks itself.

To use the Fire phone again, you must unlock it. Press the power button or the Home button, and then put your fingertip on the screen and slide up. If you've set up a PIN or password on your phone so that only someone who has the code can use it (page 317), then type the PIN or password to complete the unlocking process.

TIP While the screen is locked, the Fire phone still operates behind the scenes, checking email and Facebook on schedule. Also, you can still get phone calls and text messages while the screen is locked. You can have the Fire phone display notifications for missed calls and other incoming information on the Lock screen. Swipe down from the top of the screen or use the one-handed Swivel gesture to display the Quick Actions panel (page 18). Then tap the Settings icon (it looks like a gear) and tap Lock Screen→"Turn on notifications on the Lock screen." Move the slider from Off to On. You can also customize the Lock screen settings in other ways, too, including changing its picture (described next).

Changing the Lock Screen Image

You're not stuck with the Lock screen image you see when you first turn on your Fire phone. There are plenty of others you can use. To choose a different one, swipe down from the top of the screen (or use the one-handed Swivel gesture) to display the Quick Actions panel, select Settings→Lock Screen→"Select a Lock screen scene," and then tap the screen you want to use. The scene appears full screen. Tap the checkmark if you want to use the scene and the X if you want to choose a different scene. Tapping the X sends you back to the list of screen images so you can choose another one.

You can also choose to have the Fire phone show a different Lock screen every day, every week, or always keep the same Lock screen. To do it, pull down the Quick Actions panel (page 18), and then select Settings→Lock Screen→"Select a lock screen scene." Look toward the top of the screen for the text "Rotate Scene." Underneath it you'll see the words "Never," "Every day," or "Every week." Tap Rotate Scene, and from the screen that appears, make your choice about how often, if ever, you want the Lock screen to automatically change.

Dynamic Perspective on the Lock Screen

As you've no doubt noticed, the Lock screen uses the Fire phone's Dynamic Perspective feature (more on that on page 29) to make some remarkable-acting images. Not only is it three-dimensional, but you can also peek around and see the image from different perspectives just by moving the phone. Take a look at a few Lock screens to see Dynamic Perspective in action. A good one to try is the Barnstorming image, which shows an old-fashioned biplane in flight. Just by itself, it's pretty cool looking as the plane bounces up and down as it flies, and the clouds go rolling by underneath it. Move the phone to the left, and three more biplanes fly into view, occasionally doing rolls and tricks. Move it to the right and up—now a blimp flies into view. Keep moving the phone around for even more changing 3D perspectives.

NOTE If you turn off Dynamic Perspective (see page 30), you'll still see the Lock screens, but they'll be static images and won't change perspective as you move the phone, or have any animations.

Headset Jack

ON THE PHONE'S TOP, you'll find a 3.5-millimeter headset jack (page 5). Notice that it's a head*set* jack, not just a garden-variety head*phone* jack. It doesn't just let you listen; it accepts incoming sound as well. That's so you can plug a headset (like an earbud headset) into it and use it for making phone calls.

Of course, it's also a headphone jack, so you can plug in headphones or even external speakers and enjoy the phone as a music machine.

About the Screen

THE SCREEN IS WHERE you and the Fire phone communicate with each other. It's 4.7 inches measured diagonally. In technical terms, it has a 720 x 1280 pixel resolution and a 312-pixel density. When you turn the phone sideways, it switches to a widescreen TV and movie format.

TIP Because you're going to be touching the display with your fingers, it's going to get dirty and streaky. Simply wipe it clean with a soft, lint-free cloth or tissue. The screen is scratch-resistant, but if you're worried about scratches, get a case or a screen protector. See Appendix B for ideas.

Status Bar Icons

THE FIRE MAKES SURE to keep you updated with information about its current status and any news, updates, and information it thinks is important. It does so by displaying a variety of icons in the Status bar at the top of the screen. The Status bar is divided into two parts. On the right side, you'll find icons that indicate the Fire phone's current state, such as signal strength, 3G or 4G connection status, the time, and so on. At left is the Notification area, which alerts you when you have email or voice messages waiting, when an event on your calendar is about to occur, and so on.

NOTE Out of the box, your Status bar will be hidden, and in order to see it and the Status bar icons you'll have to use a gesture called Peek. To peek, turn your phone to a very slight angle to the left or to the right. For more details, see page 23. And to have your Status bar be always visible, without peeking, swipe or swivel to pull down the Quick Actions panel, and then select Settings→Lock Screen→"Turn on notifications on the lock screen."

Here are the most common icons you'll come across:

- **Cell signal**. The more bars you see, the stronger the signal. The stronger the signal, the clearer the call and the lower the likelihood that you'll lose a connection. If you have no connection at all, instead of this signal, you'll see the much-hated warning: "(No service)."

- **Roaming**. If you're outside your carrier's service area and connected via another network, you'll see the Roaming icon. Keep in mind that typically you're charged for making calls or using data when you're roaming, so when you see this icon, be careful what you do on your Fire phone—maybe it's not the time to download 30 songs and a half-hour TV show.

- **3G/4G**. This one appears when you're connected via 3G or 4G high-speed broadband service, which should be most of the time.

- **Bluetooth connection**. This icon indicates that you've turned on Bluetooth, for making a connection to a headset or some other device.

- **Mobile hotspot**. Your Fire phone can serve as a mobile hotspot, providing Internet service for other devices, including five computers, smartphones, or other devices and gadgets via WiFi. See page 159 for details.

- **Airplane mode**. When you use Airplane mode, you turn off WiFi and cellular communications, so you can still keep using your phone's apps, but it doesn't interfere with navigation equipment.

- **Downloading**. When you're downloading an app or media file, the downloading icon appears.

- **New email message**. You've got mail! See page 198 for more about reading new email.

- **Voicemail message**. You've got mail—voicemail, that is. (See page 72 to learn how to check your voicemail.)

- **Missed call**. Someone called you, and you didn't answer. You see this icon appear even if the person left no voicemail.

- **Time**. Shows you what time it is. Say goodbye to your watch.

- **Battery**. This shows you how much battery life you've got left. When the battery is charging, you see a battery-filling animation and a tiny lightning bolt.

- **USB connection**. The USB connection icon appears when you connect your phone to your computer via a USB cable for a variety of reasons, including copying and syncing files (Chapter 6).

Home Button

WHEREVER YOU ARE ON the Fire phone, press the Home button and you'll come back to the Home screen—either in the Carousel or Apps Grid view (page 14). When the phone is locked, press it to bring up the Lock screen.

This multitasking button has a few other tricks up its sleeve, depending on where you are when you press it:

- Pressing it in Apps Grid view brings up the Carousel, and vice versa.
- Pressing and holding it brings up voice activation features (page 82).
- Double-tapping it brings up Quick Switch multitasking mode (page 264).

Multipurpose Jack and Charger

FOR TRANSFERRING FILES BETWEEN your computer and the Fire phone, there's a micro-USB port at the phone's bottom. Plug one end of the USB cable that comes with the phone into it, and the other end into a computer, and you can transfer to your heart's content (page 141).

When your Fire phone is plugged into a computer in this way, it also gets charged, although at a slow rate. If you want to charge the phone more quickly, plug one end of the cable into the micro-USB port and the other end into the power adapter, which you plug into an electrical socket. Your Fire phone should charge in less than 5 hours. You'll most likely need to charge the Fire phone every night.

TIP You can use the Fire phone while it's charging, unless the battery has run down completely. In that case, it'll need to build up a charge before you can turn it on.

USB/Charger port

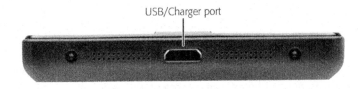

Ringer Volume

IS YOUR RINGER TOO loud? Too soft? Get it just right by using these two black buttons on the left side of the Fire phone, up toward the top.

Camera

YOUR FIRE PHONE INCLUDES not one, but two cameras that are capable of taking videos as well as photos. The camera on the back, which is the one you'll normally use for taking photos and videos, has a 13-megapixel resolution. The camera that faces you is for video calling and video chat, although you'll probably take your share of selfies as well. It's got a 2.1-megapixel resolution. To get to the camera, you can either tap the Camera icon on the Carousel or Apps Grid, or else press the small black button just below the volume buttons. That button does double-duty: Hold it down and you turn on the Fire's Firefly feature (page 42). (For more information about shooting photos and videos with these cameras, see page 116.)

The Fire phone has four other cameras as well, one at each corner of its front. These cameras don't take photos. Instead, they're used for the Fire phone's Dynamic Perspective feature (page 29).

Front-facing (selfie) camera

Four Dynamic Perspective cameras

Home Screen, Carousel, and the Apps Grid

WELCOME TO YOUR NEW home, the Home screen. Get to it by pressing the Home button no matter where you are. Unlike other phones, the Home screen has two views, not one: the Carousel view and the Apps Grid. Switch between them by pressing the Home button.

Maximizing Your Battery Charge

Your Fire phone most likely will need recharging every night. And if you use a lot of power-sucking features, you may not even be able to get through one whole day without having to recharge. In addition to turning off the screen, there's a lot you can do to make your battery last:

- **Turn off Dynamic Perspective.** It can use a lot of power. So turn it off by swiping from the top to reveal the Quick Actions panel, and then tap Settings→Display→"Configure low motion settings" and turn Dynamic Perspective off. See page 29 for details about Dynamic Perspective.

- **Turn on Low Motion Mode.** This will save even more power, because it turns off not just Dynamic Perspective but also gestures that use it, including Tilt, Swivel, Peek, and Auto-Scroll. (See page 22 for details about those gestures.) You turn it off from the same screen where you turn off Dynamic Perspective.

- **Lower your screen brightness**. Swipe from the top of the screen to reveal the Quick Actions panel, and then move the slider to a dimmer setting.

- **Turn off antennas you're not using**. If you're not using a Bluetooth headset, and don't need WiFi services at the moment, then by all means turn them off. They use up tons of power. Pull down the Quick Actions panel, and you'll find widgets for turning off (and back on) WiFi and Bluetooth. Putting the Fire phone into Airplane mode (page 162) turns off all these settings at once, as well as turning off the radio that connects you to the cellular network. You can find the Airplane mode widget on the Quick Actions panel.

- **Use headphones**. The Fire phone's speakers use up a lot of juice. Use headphones instead if you want to save power.

- **Watch out for power-sapping apps**. Some apps, such as 3D games, can run down the battery with surprising speed. If, after installing an app, you notice your battery running low quickly, consider deleting it or running it only when necessary.

The Carousel gets its name because of the way its icons rotate gracefully across the screen. In this view, apps you've recently used appear as big icons, one at a time at the top of the screen, newest first. They have no text labels, even when you use the Fire phone's Peek feature (page 23), so they can be hard to recognize the first time you see them. Also, the Carousel shows you much more than just your apps. It shows you your recently viewed books, videos, and other items. You see the most recently used item first, and as you swipe, you see apps you've previously used. You can also *pin* items to force them to show up first on the Carousel (page 16).

Underneath each big icon, you'll see the latest actions, files, or recommendations associated with that app. For example, underneath the Videos icon, you'll see recent videos you've watched, while underneath the Mail icon you'll see your most recently received emails. Tap the big icon to launch the app; tap any of the items underneath it to launch that content in the app.

Down at the bottom of the screen in Carousel view you'll find four icons:

- **Phone** for making phone calls.
- **Messaging** for sending and receiving text messages.
- **Email** for sending and receiving email.
- **Silk Browser** for browsing the Web.

The Apps Grid displays all of your apps in a scrolling grid. Keep swiping up to get to more apps until you come to the end. Tap any icon to launch the app.

Customizing the Carousel and Apps Grid

You're not stuck with what's in the Carousel; you're free to remove items from it. And you can add items to it as well. To remove an item from the Carousel, hold your finger on it and choose "Remove from Carousel." If you have a favorite app or other item, you can make it appear first in the Carousel—choose "Pin to Front." If you hold your finger on an app that can be uninstalled from the Fire phone, a third choice will show up—"Remove from Device."

NOTE Not all apps can be removed. You can't, for example, remove built-in apps such as Email and Videos.

You can remove apps from the Apps Grid in the same way: hold your finger on any and select "Remove from Device." You can also change the order in which apps appear by dragging an app from one location to another.

Left and Right Panels

IN ANY APP OR screen on the Fire phone, you can display left and right panels, which give you more features and options related to whatever you're currently doing. To display the left panel, swipe from the left side of the phone toward the center of the screen, and to display the right panel, swipe from the right side. You can also use the one-handed Tilt gesture (page 26) to display them. To open the left panel, tilt the top of the phone to the right; to open the right panel, tilt to the left.

When you're on the Carousel or Apps Grid, the left panel displays a list of various types of content you can browse, including Apps, Games, Music, Books, and so on. The right panel shows your most recent appointment from your calendar, the local weather, and recent text messages. You'll see different information in these panels depending on what you're doing. The left panel typically displays additional menus and offers navigation features—for example, when you're reading a book, you'll see its table of contents in the left panel. The right panel typically shows information or commands relevant to what you're currently doing. For example, when you're listening to music, the right panel uses X-Ray to show you the song's lyrics (page 131), and when you're writing a text message, the right panel lets you attach a photo.

Quick Actions Panel

WHATEVER SCREEN YOU'RE ON, you can swipe down from the top of the screen to display the Quick Actions panel. You can also use the one-handed Swivel gesture (page 27) to display it. Just swivel your wrist to either the left or right. As the name suggests, this panel gives you immediate access to common actions—like changing settings, turning on WiFi, and so on. Here are your options:

Tilt phone slightly to the right or left to reveal labels

- **Airplane mode** turns on and off all of your Fire phone's radios, including WiFi, cell service, and Bluetooth.

- **WiFi** turns WiFi on and off (page 154).

- **Bluetooth** turns Bluetooth on and off (page 74).

- **Flashlight** turns your Fire phone into a surprisingly powerful flashlight by turning on the camera's flash and keeping it on until you tap the Flashlight icon again.

- **Sync** tells the Fire phone to contact Amazon and grab any new purchases, information, and updates. For example, syncing downloads items you've recently bought from Amazon, such as books and videos. It also gets you any software updates that are available for the Fire phone and downloads any new or updated apps.

- **Settings** gives you access to all the Fire phone's settings (page 299).

- **Mayday** puts you in touch with someone at Amazon to get tech help (page 36).

- **Search** finds any terms you type in. It's a big search—it searches content on your device, in the cloud, in the Amazon stores, and on the Web.

Underneath these icons, there's a slider that lets you make the screen brighter or dimmer. If you prefer, tap Auto to have the phone detect ambient light and choose the best brightness setting.

Underneath that are notifications of all kinds—messages you've received, apps updated, and more.

The Software Behind the Fire phone

I hear that the Fire phone is an Android phone. It kind of looks like one, but not completely. Is it Android?

The short answer: Yes and no.

Here's the long answer:

Buried deep in the Fire phone is Google's Android operating system, which powers many other popular smartphones. (Basically, any popular phone that isn't an iPhone, Windows phone, or BlackBerry is probably an Android phone of one kind or other.) But Amazon has dramatically changed Android in the Fire phone and called it the Fire operating system. At this writing, the Fire phone uses Fire OS version 3.5.1 (but you'll get updates, as discussed on page 287). Fire OS lets the phone do plenty of things that Android can't do, like using Dynamic Perspective and one-handed gestures.

Still, you'll find lots of Android underneath, notably when you're using individual apps, since the apps you'll find in the Amazon Appstore (page 259) are actually Android apps that have been tweaked to run on the Fire phone. But because the Fire phone doesn't run true-blue Android, there are things that it can't do, such as run Google Maps or the Android YouTube app. In some instances, the Fire phone operating system uses features similar to but different than Android's, for example, its voice control features (page 82).

Controlling the Fire with Your Fingers

WITH THE FIRE PHONE, your fingers do the walking. They do all the work that you do on a computer with a mouse or keyboard. Here are the basic finger strokes you can use on the phone's screen.

NOTE The Fire phone also makes use of *one-handed* gestures, which let you do all kinds of navigation tricks with just a single hand. For details, see page 22.

Tap

Tapping is as basic to the Fire phone as clicking is to a mouse. This simple gesture is how you press onscreen buttons, place the cursor for text entry, and choose from menus.

Touch and Hold

Touch an object and hold it for several seconds, and depending on what you're holding, an option menu appears. For example, when you touch and hold an icon on the Apps Grid, a menu appears that lets you delete the app. You also touch and hold an object to grab it and drag it somewhere.

Drag

After you've grabbed something, you can drag it with your finger—like dragging an icon to the Trash.

Swipe

Swipe your finger across the screen to perform some specific tasks, like unlocking your phone after it's been put into Standby mode, or answering a call if the phone is locked.

Flick

Think of the flick as a faster slide, done vertically when scrolling through a list, like your contacts list. The faster you make the flicking motion, the faster your screen scrolls—sometimes too fast. You can stop the motion, though, by touching the screen again. To scroll through large lists quickly, you can flick multiple times. (You can also scroll in some apps using Dynamic Perspective by merely tilting the phone. See page 28 for details.)

NOTE Flicks seem to actually obey the laws of physics, or at least as much as virtual movement can. When you flick a list, it starts off scrolling very quickly, and then gradually slows down, as if it were a ball set in motion that gradually loses momentum.

Pinch and Spread

In many apps, such as Maps, you can zoom in by spreading your fingers—placing your thumb and forefinger on the screen and spreading them apart. The amount you spread your fingers will determine the amount you zoom in.

To zoom out, put your thumb and forefinger on the screen and pinch them together. The more you pinch, the more you zoom out.

Double-Tap

When you're viewing a map, a picture, or a web page, you can zoom in by double-tapping. In some instances, once you've reached the limit of zooming in, double-tapping again restores the zoom to its original size. In Maps, you can also zoom out by tapping with two fingers.

Go Back

At first glance the Fire phone seems to be missing something important—some kind of back button that lets you return to the screen you were on previously. It's true that the Fire phone has no back button, but that's because you don't need one. Instead, you use a gesture to go back to where you once belonged.

To do it, when you're holding the phone in portrait mode (that is, vertically), swipe up from the bottom of the screen. Voilà! There you are, back to the screen you were on last. If you're holding the phone in landscape mode (that is, horizontally), and are running an app that runs in landscape mode, then you'll have to swipe from the *side*.

NOTE Be careful when making the Go Back gesture that your finger starts below the bottom of the screen. It will be to the left or right of the Start button, not on the Start button itself.

When you use go back on the Lock screen, you'll go to the last screen you were using before the Lock screen turned on. However, the Go Back gesture doesn't have a long memory. So if you turn your phone off and then back on again, it won't remember which screen you were on when you turned off your phone, and you'll end up on the Home screen.

One-Handed Gestures

WITH THE HELP OF the Fire phone's four front-facing Dynamic Perspective lenses (page 13), you perform many tasks with just one hand—the one you're holding the phone in. One-handed gestures are not only convenient; they also let you do things that you can't accomplish with the standard tapping and swiping gestures.

TIP If you find one-handed gestures difficult to use (or just plain don't like them), you can turn them off. Swipe from the top to pull down the Quick Actions panel and select Settings→Device→Manage Accessibility→Low Motion.

Peek

Turn your phone to a very slight angle to the left or to the right, and you'll see otherwise-hidden pieces of information about whatever is on your screen. For example, if you're using Maps and searching for "clubs," peeking reveals information about each pinned club on the map, including its name and Yelp rating info.

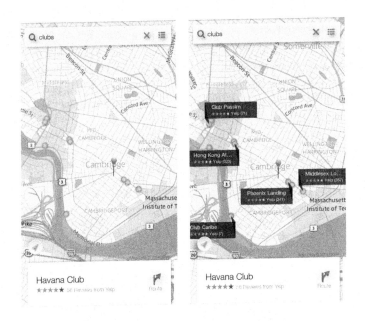

Peek also lets you peek around landmarks in Maps, giving you different 3D views of landmarks and buildings, and even looking behind and around them.

One-Handed Cheat Sheet

Since you've never used one-handed gestures on any phone you've owned so far, it's easy to forget to use them at all. Or you may trigger them inadvertently with a random jerk of your hand. (If that happens, just swipe up from the bottom to return to whatever you were doing.)

Here are the most common uses for each of the gestures. These are the ones you should get used to using:

- **Peek.** Peeking does things like reveal labels for seemingly nameless icons. It usually works in Amazon apps and screens, like the Quick Actions panel and Photos. It's also great to use in Maps (page 248).

- **Tilt.** By far, tilting is used more for opening the right and left panels than anything else. It works almost everywhere on the Fire phone, especially when you're working with music, photos, videos, and books.

- **Swivel.** Swiveling is the one-handed way to open the Quick Actions panel (page 18).

- **Auto-Scroll.** This one has a lot of potential, but right now the only place you'll probably use it is in the Silk web browser (page 171), to scroll down long web pages.

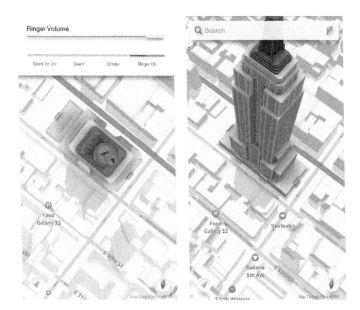

If you're in the Quick Actions panel, then when you peek, you'll see the name of each icon.

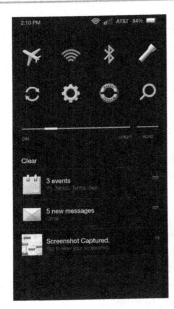

If you're shopping in the Kindle Store, then when you peek, you'll see ratings for books.

Peeking works on just about every single screen on the Fire phone. Sometimes the effect will be so subtle you won't even notice it—for example, it may do nothing more than make menu items appear three-dimensional. You can make up a new parlor game—what will peeking do on which Fire phone screen?

Tilt

Get to know this gesture, because it may be the one that you'll find yourself using most often. Tilt is the one-handed way to open the ever-useful left and right panels (page 17). To open the left panel, tilt the phone's left edge toward you and then back. This gesture works best if you do it in one sharp, swift motion. If you do it slowly or don't bring it back to the start position, you may instead tell your Fire phone to peek instead of tilt.

You can also use the Tilt gesture to make the panel disappear. Simply reverse the motion: tilt the phone's left edge away from you and then back toward the position in which it started.

To open the right panel, do the same thing from the other way: tilt the phone's right edge toward you and then back. And to make the panel disappear, reverse the motion.

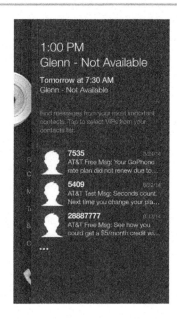

Swivel

While you're holding the phone in portrait mode—that is, vertically—swivel your wrist either to the right or to the left and you'll open the Quick Actions panel. In other words, you're swiveling either the top-right or top-left corner of the phone left or right.

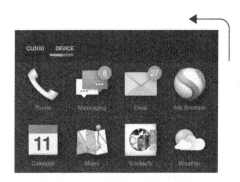 

The Quick Actions panel is a fast, convenient way to perform all kinds of common tasks, like putting your phone into Airplane mode, using it as a flashlight, connecting to a WiFi hotspot, turning on Bluetooth, searching the phone, getting Mayday support, and much more. For details, flip back to page 18.

Not all one-handed gestures work when you're on the Lock screen. For example, if you try using Tilt, you'll instead *peek* around the Lock screen. On the other hand, Swivel works on the Lock screen (and everywhere else) to display the Quick Actions panel.

Auto-Scroll

This gesture lets you scroll through long web pages on the Fire phone's Silk browser, as well as through pages or screens on any other app that supports the gesture. To use it, angle the phone away from you to scroll down. An arrow appears toward the bottom of the screen to show the direction of your scroll. Angle the phone toward you to scroll up. Again, an arrow appears toward the top of the screen to show the direction of your scroll. You can change the pace of scrolling—the more you tilt the phone, the faster it scrolls.

Auto-Scroll works only with apps that have been built to use it. For example, when the Fire phone was first released, Auto-Scroll didn't work with the Kindle app. Amazon is working on it, though, and says the Auto-Scroll Kindle feature will include a lock icon that will scroll at the pace you set without your having to keep tilting. That's a good thing, because it would be pretty tough to read a book while you're simultaneously tilting the phone toward and away from yourself.

Dynamic Perspective

ONE OF THE FIRE phone's unique features is Dynamic Perspective, which lets you take some kind of action by merely tilting the phone. For example, when you're viewing a website in Silk, you can tilt the phone to scroll. Tilt it slightly and you scroll through web pages slowly; tilt it more extremely and you scroll faster. In Maps, you can get a full 3D view around a building merely by tilting the phone at various angles (page 248).

NOTE   Dynamic Perspective also powers one-handed gestures, like peeking behind an object, or revealing the left or right panel, as described in the previous section.

How Dynamic Perspective Works

The Fire phone uses some pretty impressive hardware to make Dynamic Perspective work its magic. Four—count 'em, four—cameras on the front of the phone are devoted to this feature. Their lenses are easy to see; there's one on each of the camera's four front corners. And they're not ordinary cameras. They're infrared cameras, which lets Dynamic Perspective work even in the dark.

The software underlying it all has been optimized for recognizing human faces, so it recognizes faces and the angle they're looking at the phone. The software uses the cameras as sensors to determine your position relative to the image that's being shown on the screen. One of its best features is that it lets you "peek" behind and around objects to see what's behind them, or to see them from different angles as you move your phone in relationship to your face. Here are just a few of the things you can do with it:

- Look at the Empire State Building from different angles by merely moving the phone.

- Look around and behind 3D objects on your Lock screen.

- If you're looking for seats at a sports stadium, arena, or other venue in an app like StubHub, you can tilt the phone and see the actual view you'll get from your seat in multiple directions, just as if you were actually there.

- You can use Dynamic Perspective as a kind of game controller (page 32) and can also look around corners and obstacles in games before you get there, so you're not surprised by any nasties.

There's more as well, because Dynamic Perspective is built into the very DNA of the Fire's operating system. With time, more and more apps will take advantage of its powers. Amazon has publicly released a Dynamic Perspective software development kit (SDK) and an associated application programming interface (API). That means that independent developers can build all kinds of Dynamic Perspective capabilities right into their apps. Expect plenty of them in the future.

NOTE Unfortunately, Dynamic Perspective can eat up battery life. See the box on page 14 for more detail.

Customizing Dynamic Perspective

You might not like certain one-handed gestures, or you might not like Dynamic Perspective at all and want it turned off entirely. To do it, swipe or swivel to pull down the Quick Actions panel and select Settings→Device→Manage Accessibility→Low Motion. To turn off Dynamic Motion entirely, including all one-handed gestures, turn Low Motion Mode at the top of the screen from Off to On.

Is Dynamic Perspective Really Always Watching You?

Dynamic Perspective works only when the phone can see your face. But how can you be sure of that? It's a bit like the famous conundrum, "If a tree falls in the forest and nobody hears it, does it make a sound?" However, in this case, you can actually prove that if your Fire phone doesn't see your face, then Dynamic Perspective won't spring into action. Try this: First, turn on your phone. Look at the Lock screen. See how it changes perspective? OK, now you're sure it's working.

Now hold your phone sideways and away from you so that the cameras can't see your face. Once you've done that, move the phone around in a gesture that Dynamic Perspective would normally recognize. Make sure that the phone can't see you but you can still see the screen. If you do it right, you'll see that the Lock screen doesn't move as it would normally, because it can't see your face. (As for the tree falling in the forest, Dynamic Perspective has no answer for you about that.)

If you want to instead turn off only one or several one-handed gestures, don't turn Low Motion Mode on. Instead, turn off any one-handed gesture you don't want to work anymore. To change the settings back to the way they were, come back to this screen.

Using Dynamic Perspective as a Game Controller

To get a sense of what Dynamic Perspective can really do, try a few games that use Dynamic Perspective. Some of them even use it as a game controller. It's clearly the future of immersive 3D gaming. There are a number of games that use it that you can try out now—search for them in the Amazon Appstore (page 259). For example, the free Saber's Edge is a combination fighting and strategic 3D puzzle game. It uses Dynamic Perspective to peek around corners of a puzzle's cube, or to tilt the cube left or right to another face. If you're more interested in action and using your head as a game controller, try the free Snow Spin. Just move your head left or right to control the snowboarder.

NOTE Not all games can use Dynamic Perspective—only those whose developers have specifically built it in can make use of it. Apps that use Dynamic Perspective say so on their info pages in the Appstore.

Amazon Prime

ONE OF THE FIRE phone's biggest benefits is a free year of Amazon Prime. This set of premium services normally costs $99 a year. That may well exceed the cost of the Fire—at one point Amazon dropped the price to 99 cents (page xvi). At that price, if you were already planning to subscribe to Amazon Prime, you essentially got your phone and Amazon Prime for free for a year. That's a pretty good deal, and one that no other phone can offer.

NOTE If you already have a Prime subscription, you get an additional year tacked onto the end when you buy a Fire Phone.

When you register your phone during your initial setup process, make sure to use your existing Amazon account if you have one. That way, that account will get all the Prime benefits. If you don't already have an Amazon account, you'll create one when you buy and register your phone.

Free Shipping

With Amazon Prime, you get free shipping, and you don't need to order from your Fire phone to take advantage of it. You'll get free two- or next-day shipping or a reduced price on same-day shipping on millions of eligible items when you buy on the Web or from another device. (The exact shipping charges vary based on where you live.)

NOTE Some of the items you buy on Amazon are sold not to you directly by Amazon, but instead by a partner vendor. Free Prime shipping applies to some of these vendors but not others. If you see the Prime logo on the item's info page, shipping is free.

Prime Instant Video

With Prime Instant Video, you get to watch an impressive video collection, including movies, TV shows, and Amazon-created series not available anywhere else, such as the political comedy *Alpha House* starring John Goodman and written by *Doonesbury* creator Garry Trudeau. Any video that's labelled as "Prime Instant Video," you can watch for free. Any videos that aren't labelled "Prime Instant Video," you'll have to pay to either rent or buy.

In addition to watching the videos on your Fire phone, you can watch it on other devices like a Kindle, iPhone or iPad; a TV with an Internet connection or Fire TV or Fire Stick TV; a Blu-ray player; a gaming console; and almost any computer with an Internet connection.

On your Fire phone, you watch Prime Instant Video via the Instant Video app that comes with the phone (page 121).

Prime Music

An Amazon Prime subscription includes free access to Prime Music, which lets you stream ad-free music to your Fire phone, Kindle Fire, PC or Mac, or other device. You select individual songs and albums and create playlists, and you can also listen to other people's playlists. You get access to over a million tracks. Download the Prime Music app (page 133) to your Fire phone to listen to music there.

When an album is on Prime Music, sometimes you'll find that not all tracks are available. And keep in mind that while a million tracks sounds like a lot, it's still a much smaller collection than the 20 million on Spotify (page 136).

Access to Free Books

With Amazon Prime, you can read plenty of books for free, via two different Amazon programs—the Kindle Owners' Lending Library and Kindle First.

The Lending Library gives you access to 500,000 titles. You can borrow and read the books for as long as you want; there's no due date. You can read the books not only on your Fire phone, but on any Kindle device registered to your Amazon account. You can borrow one book a month. In the Kindle Store (page 97) swipe to reveal the left panel, and then scroll down and tap Kindle Lending Library. When you find a book you want to read, tap the "Borrow for Free" button. For more details on borrowing books, see page 99.

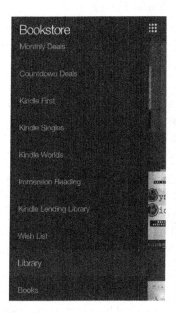 

With Kindle First, you can read one of four Kindle books—chosen by Amazon editors—before they're published. Swipe to reveal the left panel, scroll to Kindle First, and then select a free book. Feel special about getting to read it before everybody else. The books change every month.

Mayday

WHO DOESN'T HATE TECH support? Telephone trees that make you wish you could uproot them, long waits on hold while listening to the kind of music that makes you long for a lobotomy, all to get through to a probably clueless tech support rep.

Amazon's Mayday feature changes all that. It lets you get live tech support right on your Fire phone. You can see and talk to your tech-support person in real time. These Tech Advisors can draw on your screen to show you what to do, or even take control of your phone and fix the problem right then and there. Don't worry if you're asking for tech support in your pajamas—the Tech Advisor can't see you.

NOTE Where does the name "Mayday" come from? It has nothing to do with workers' holidays or dancing around poles. The word "mayday" signals a distress call from a ship or airplane. It originated from the French term for "help me"—*m'aidez,* which is pronounced much the same way.

Pull down the Quick Actions panel and tap the Mayday button. From the screen that appears, tap Connect.

Very soon an Amazon Tech Advisor appears in a video window at the bottom of your screen. Amazon says the wait is usually less than 15 seconds, and people often find that to be the case, although it may take up to 30 seconds. (Sometimes you may not be able to see the tech advisor because of connection issues. In that case, you'll instead see an icon of a headphone to show there's no video connection.) Start talking and explain what you need help with. More often than not, a few rounds of questions and answers will solve your problem. (Yes, such kinds of interactions still exist. They're called *conversations*.) The Tech Advisor can see your screen, so you can go to where you're having the problem while you speak.

Again, if the Tech Advisor can't describe the solution in a way you can understand, he can draw right on your screen and show you what to do. For example, say you don't know how to play your music from the cloud in the cloud player. After you explain that to the Tech Advisor, he may draw a circle around the Music app and tell you to tap it to launch it.

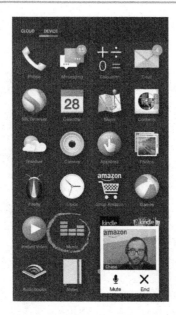

Then, when the Music app launches to the Album view, he may draw an arrow on the screen to show you that you should swipe from the left in order to display the left panel.

When the left panel appears, he may circle the word "Cloud" at the top and tell you to tap there. That will solve the problem.

TIP If you don't get something fixed the first time, ask the Tech Advisor to repeat it. Even if they don't know the answer immediately, Amazon's Tech Advisors have for the most part proved to be patient and willing to make every effort to be helpful.

When the problem is solved (or even if it isn't and you just want to quit), tap the End button below the Tech Advisor to end the session. A screen will appear asking whether your problem was solved.

If that doesn't work, there's an even more sophisticated solution. The Tech Advisor can operate your Fire via remote control and show you all the steps you need to take to do something. That way you can see it in action.

NOTE The Tech Advisor may not always be able to establish a connection with your phone that lets her draw on your phone or operate your Fire via remote control. In that case, she'll have to talk you through the solution.

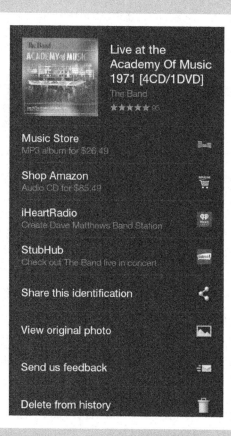

Live at the
Academy Of Music
1971 [4CD/1DVD]
The Band
★★★★★ 95

Music Store
MP3 album for $26.49

Shop Amazon
Audio CD for $85.49

iHeartRadio
Create Dave Matthews Band Station

StubHub
Check out The Band live in concert

Share this identification

View original photo

Send us feedback

Delete from history

You'll learn to:

- Learn what Firefly can and can't recognize

- Use Firefly to recognize and buy books and music

- Use Firefly to identify and rent or buy movies and TV shows

- Connect with phone numbers, email addresses, and websites by pointing your Fire phone at them

CHAPTER 2

Firefly

FIREFLY SOMETIMES SEEMS MAGICAL. With this technology, your Fire phone looks at things you point it at—text, posters, artwork, phone numbers, email addresses, and more—and identifies them. Firefly can listen as well, and recognizes music, movies, and TV shows by what it hears. If the item is available from Amazon, Firefly lets you buy it with a tap.

Understanding Firefly

FIREFLY RECOGNIZES SO MANY things that when you first try it you may find yourself pointing it at random objects just to see what it comes up with—like a nearby container of Lysol wipes. Sometimes it recognizes them and other times not. But it's always entertaining to see what it finds.

When you point Firefly at an object, it does some fancy object-recognition and matching (see the box on page 43) and compares what it finds to images in Amazon's product database. If Firefly can't figure out what you're pointing it at, try pointing it at a bar code. Firefly can recognize just about anything that way.

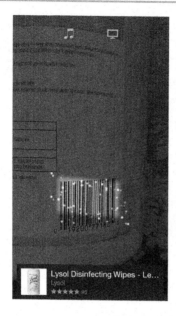

When it comes to images, Firefly can do amazing things. For example, point it at a playbill for a community theater musical play, and it can recognize the show and give you options for buying the cast recording for the Broadway show and creating a music station based on it in iHeartRadio (page 51).

If you try a banana, a pear, or a peach, however, you may not have any luck.

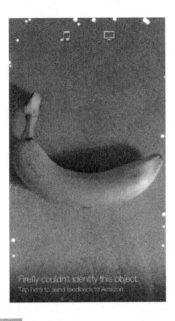

Firefly couldn't identify this object.
Tap here to send feedback to Amazon.

How Firefly Works

Firefly uses some very impressive image recognition and matching to identify objects. It starts off by scanning an object and identifies parts of it that are uniquely identifiable. For example, when scanning a box of cereal, it concentrates on the outline and color of the box, and on the logo. Firefly will adjust how many points and pieces of data it looks at, depending on the object it is trying to identify. For example, it uses 1,000 different points to get enough data to recognize a painting.

Firefly takes just the shapes' outlines and other information such as color—not the full image itself—and sends them to Amazon cloud servers. The servers then look for a match through Amazon's massive product catalog.

The database has millions of objects in it, and it would be too difficult and time-consuming to try to match the information about the box against every object in it individually. So instead, Amazon's image-matching algorithm tries to match only specific attributes of the scanned image, such as the shape of the box, the color of the box, and the logo, against objects with similarities.

The image recognition generally works quite well, although you may sometimes find surprising results. For example, if you point your Fire phone at an image of a Monet painting inside a book, Firefly may find the matching object not to be an actual painting, but instead a decorative iPhone cover that has a picture of the painting on it.

Getting Started with Firefly

THERE ARE TWO WAYS to launch Firefly:

• Tap its icon in the Carousel or Apps Grid ![icon]

• Press and hold the Firefly button on the bottom left of the Fire—it also serves as the camera button (page 12).

> **NOTE** Make sure to press *and hold* the camera button. If you only press it, you'll instead launch the camera.

Once you've launched Firefly, point the camera at a product to scan and recognize it, or hold your Fire near an audio source in order to recognize music, TV, or movies. You'll see little fireflies floating around on the screen. That means Firefly is "thinking."

If your camera is pointing at a physical object, all you have to do is wait until Firefly works its magic. To recognize music, TV, or movies, you have to take an additional step: Tap the music icon to tell Firefly that you want to identify music; tap the video icon to recognize TV shows or movies instead.

There are some things you can do to help Firefly best recognize what you want it to. Hold your Fire phone close to the object you want it to identify and make sure you hold it steady. Tap the phone's screen—that will help Firefly better focus on the object. And if Firefly can't identify the object by what's on the front, look for a bar code and point Firefly at that. It'll identify the object fast.

If Firefly can identify the object, you'll see a label at the bottom of the screen. What you see varies according to what you're identifying. For a book, for example, you'll see the name of the book and its user rating. Tap the label to get more information about the object and to buy it via Amazon.

Just How Smart Is Firefly?

How many things can Firefly recognize? If you believe Amazon, it's more than 100 million. That includes 70 million products of various kinds, 245,000 movies and TV episodes, 160 live TV channels, and 35 million songs. As the world creates more products, music, movies, and TV shows every day, that number will continue to grow.

That's pretty smart, but it doesn't include every object in the world, or even every object for sale. It matches only objects that are in Amazon's product database, because Firefly's main purpose is to let you buy things you see through Amazon.

Using Firefly with Books

GIVEN THAT AMAZON STARTED as a book-selling site, and it's the largest online book-selling site in the world, it's no surprise that Firefly does a stellar job of recognizing books, giving you information about them, and, of course, making it easy for you to buy them.

Point Firefly at a book's front cover, and most of the time, Firefly will recognize it. If not, though, turn the book over and point it at the bar code, and Firefly will almost certainly recognize what it is.

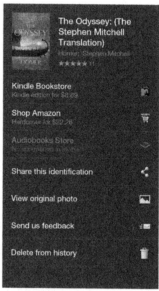

Keep in mind that you can do this not just for books that are currently in print, but for out-of-print books as well. However, older out-of-print books may not have reader ratings.

Out-of-print books do not include
buying and related information

Now that you've recognized a book, what's next? Tap the label, and you'll come to a screen with the following options:

- **Kindle Bookstore.** Tap here to buy the Kindle version of the book. You can also usually download a sample of the book, read reviews, see related books, and so on.

- **Shop Amazon.** Tap here to buy the book on Amazon—the print book, that is, not the Kindle book. You'll see either the hardcover or paperback. It's not clear how Amazon decides what to put there, but it's likely whichever is selling more on Amazon. This page also shows where you can buy the book, read reviews, and so on.

- **Audiobooks Store.** Tap here to buy the audio version of the book if one is available.

- **Share this identification.** Share the book with others, via Bluetooth, email, Facebook, Twitter, and other methods. It sends the title of the book as well as a link to it on Amazon and the text "Shared from Firefly."

- **Send us feedback.** Submit comments about Firefly. The feedback is sent to Amazon's Firefly team. Depending on the comment, it may be used to help the Firefly team analyze the image and improve accuracy in the future.

- **Delete from history.** Lets you delete the book from your Firefly history list. For more details about managing your history, see page 56.

Using Firefly with Music

FIREFLY CAN RECOGNIZE MUSIC in two different ways: the easy way, and the hard-to-believe, mind-boggling way. The easy way: Point your camera at the front of a CD case or at the bar code on its back, just as if it were a book, and Firefly recognizes the CD. What if you're old-school and prefer vinyl LPs to CDs? No problem; Firefly can recognize their covers as well.

Whether it's a vinyl LP or CD, after Firefly recognizes the music, it displays a label like the one it does for books. Tap it to get more details.

Now here's the amazing way: If you're listening to music, point Firefly at the music source and tap the music icon. Firefly listens, and then identifies the music. When you do it this way, Firefly may skip displaying the label and send you straight to the information about the track.

There's plenty you can tap on this screen:

- **Music Store.** Buy the MP3 track from Amazon's music store so you can play it on your Fire phone (page 127).

- **Shop Amazon.** Buy the physical CD from Amazon.

- **StubHub.** Tap this option, and you'll see a list of the artist's touring schedule near where you live. For more details about any event, and to buy tickets, tap the event. (You need a StubHub account and the StubHub app from Amazon's Appstore to use this ticket-buying service with your Fire phone.)

- **iHeartRadio.** Lets you create a custom station of the artist's music—and music from similar artists—on the iHeartRadio service. As with StubHub, you'll need to first download the iHeartRadio app from the Amazon Appstore and register. Once you do that, iHeartRadio creates and launches a custom station with just a tap. (For more details, see page 277.)

- **Share this identification, Send us feedback, Delete from history.** These options behave exactly the same way as for books recognized by Firefly (page 47).

Using Firefly with Movies and TV Shows

FIREFLY EXCELS AT RECOGNIZING movies and TV shows. If you point it at a DVD cover, it does a great job of figuring out what it's looking at, even if it's not a well-known production. And just as it can recognize old vinyl LPs, it can recognize old VHS tapes by the packaging as well.

As with music, Firefly can identify TV shows and movies merely by listening to the audio. But to help Firefly know what it's supposed to be listening to, you have to tap the video icon first.

TIP If Firefly doesn't recognize the movie or TV show right away, don't despair. The second or third try often works better.

Finally—and most remarkably—if you point Firefly at a screen that's playing a movie or TV show, Firefly may recognize it that way as well. Again, tap the video icon first.

No matter how Firefly does the recognition, when you tap the label at the bottom of the phone's screen, you come to a page that gives you a variety of possible actions—buying the DVD on Amazon, sharing a buying link for it with others, and so on. If the movie or TV show is in Amazon's Instant Video library (page 121), you'll get a wealth of details about it, including which actors are in the current scene, how far along the episode is, and more. And you can, of course, also watch it on Amazon Instant Video on the Fire phone, Kindle Fire, Fire TV, or computer.

If you install the IMDb app on your Fire phone, Firefly integrates with it as well. You can simply tap to see a wealth of details on the Internet Movie Database (which is, not surprisingly, owned by Amazon).

Using Firefly with Phone Numbers, URLs, and Email Addresses

FIREFLY DOESN'T JUST RECOGNIZE products by their packaging, audio, or video. It can also recognize text and numbers and figure out whether they comprise a phone number, email address, or web address. (In that respect, Firefly may be smarter than some people you know.) Then it hooks right into the appropriate app on your Fire phone—the Silk browser—so you can visit a web page, for example. Point Firefly to a URL on your computer display, and then tap the label to come to a page where you can tap to go to the website, view an image that the URL points to, and so on.

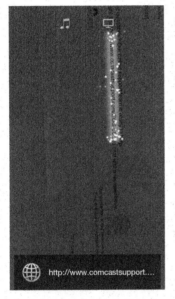
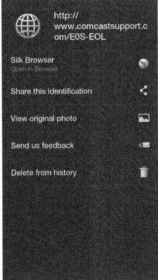

When Firefly recognizes a phone number, it shows you the number, along with an icon of a phone next to it. Tap the icon and you'll get a list of choices, including dialing the number, adding the number to your contacts, sending a text message to the number, and so on.

When it recognizes an email address, it will do the same thing, and offer options such as addressing an email to it. And if it recognizes several numbers at once—for example, a phone number and an email address—it will give you options for acting on both of them.

Firefly sometimes has problems recognizing email addresses and phone numbers. Moving to different lighting conditions often solves the problem. Sometimes brighter conditions help, and sometimes darker—give one a try if the other doesn't work.

What to Do with Your Firefly History

FIREFLY KEEPS TRACK OF every item you've identified with it. That's more useful than it sounds. For example, if you want to revisit an item—get details about a movie you identified a week ago—you can do it in a flash.

When you're in Firefly, swipe up from the bottom of the screen, and your history list appears. Scroll through it, and then tap the item that you want details about.

You can search through the list as well, which is useful if you use Firefly a lot. As you type, the results keep narrowing down. Keep typing until you find the result you want, and then tap it.

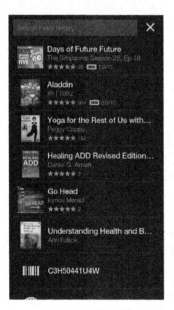 

Once you start using Firefly, you may find it addictive, so you can quickly build up a long history list. It may get so long that it's almost useless, since you'll have to spend too much time scrolling or searching through it to find old items. If you want to delete an item, tap it, and then choose "Delete from History."

NOTE Firefly keeps track of your history whether you like it or not. There's no way to turn it off.

Incoming Call...

Preston Gralla

Author
Work

Answer
Hold current call

Answer
End current call

Decline

You'll learn to:

- Make calls using the phone dialer, Contacts, and VIPs

- Use special features like conference calling, call waiting, and visual voicemail

- Send and receive text messages

- Control the Fire phone with your voice

CHAPTER 3

Phone Calls, Text Messages, and Voice Control

THE FIRE PHONE HAS so many unique skills, you may forget it can handle phone calls and text messages as well. But the Fire phone offers a full gamut of calling features—everything from simple voice calls to nifty features like Call Forwarding. And it lets you embellish your text messages with pictures and video, too.

Whenever you see by the bars in the screen's upper-right corner that you've got cellular reception, you're ready to make calls. You can place a call in any of five ways. Don't be daunted by the variety—all methods are easy, including a "Look, Ma, no hands" approach to calls that lets you call someone by talking into the phone rather than by using your fingers to tap keys.

Using the Phone App

THE FIRE'S PHONE APP is Command Central for making phone calls. On the Carousel, tap the phone icon (at the lower left). You can also tap the icon on the Apps Grid. The Phone app opens, with four buttons at the bottom:

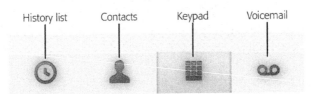

History list   Contacts   Keypad   Voicemail

- **History list**. Shows all the calls made to you and that you've made to others. Tap any of the listings, and you'll call that person.

- **Contacts**. Tap a contact, and you can choose from among making a call, sending a text message, sending an email, and more. The options you see here depend on the information you have for that contact. If you don't have someone's email address, for example, you won't get that option. You'll find out more about all these options on page 62.

- **Keypad**. You'll be pleased to see that the virtual buttons on this dialer are nice and big. Even if you have fat fingers or iffy coordination, tapping the right number is a breeze. Tap the number you want to call, and then tap the blue Call button at the bottom.

- **Voicemail**. Tap here to check your voicemail.

Placing a Call

MAKING A CALL WITH the dialer is simple: Tap the virtual buttons, and then tap the blue Call button to place the call.

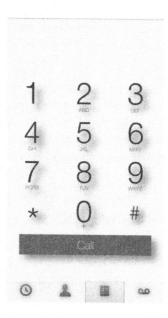

If you want to call in to your voice mail instead (see page 72), tap the voicemail button at the lower right.

When you make a call, the Dialing screen pops up showing you the phone number. If you have a picture of the person in your Contacts list, you see that photo here. A timer begins, showing you the elapsed time of the call. There are also buttons for putting the call on the Fire phone's built-in speaker, muting the call, calling up the keypad during the call, and a + button, which lets you conference in another person. (For details, see page 67.) And, of course, to end the call, tap the End Call button.

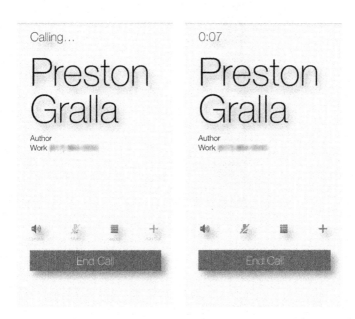

That Keypad button may seem baffling. Why would you need a keypad when you're already on a call? For that most annoying part of modern communication—the phone tree. Press the Keypad button, and you can experience all the joys of having to "Press 1 for more options."

Choosing from the History List

The history list shows you every call you've made or received and lets you make phone calls right from the list. Tap the History button at the bottom of the screen—it looks like a clock—and you see your call history listed in chronological order. Tap any to make a call. To delete a listing, hold your finger on it, and then select Delete Call.

NOTE The number in the history list's icon shows you the number of calls you've missed.

Number of
missed calls

Choosing from Your Contacts

The Phone app gets phone numbers, pictures, and other information by tapping into your main Contacts list (page 108). The Fire phone gets these contacts from multiple places. For example, if you use Gmail, the phone automatically imports your Gmail contacts into its list. And if you use the Fire phone along with Facebook, your Facebook friends join the party as well. You can also add contacts directly to the phone. Suffice it to say, if you've ever interacted with someone by email or social media, chances are your Fire phone already has his phone number.

NOTE You may notice that on your Contacts list, many people have pictures next to their names, even though you never took their pictures. Is the Fire phone pulling hidden camera tricks? Of course not. If you use Facebook on your phone (page 221), it pulls in profile pictures from Facebook for any of your contacts who have them.

You likely know more than one screenful of people, so you can navigate through the list in several ways. First, you can flick through the list. You can also search the list. Type letters into the search box at top, and as you type, the list gets pared down, hiding everyone whose first name, last name, company name, or title doesn't match what you've typed. It's a great time-saver for quickly paring down a big list. You also see the words you're typing, just above the keyboard, so you can more easily track what you type.

TIP How do you find a contact whose name you can't recall? If you know her place of work, type that in, and you'll see a list of all your contacts who work at that company.

When you've gotten to the person you want to call, tap the person's contact listing. You'll see all the information you have about him: phone numbers, email addresses, home and work addresses, and more. From this screen you can make a call—or send an email or a text message—by tapping the appropriate icon.

Using the VIP List

WHAT IF YOU WANT to be able to quickly call your best friend, spouse, law-yer, weekly tennis partner, or kids? Scrolling through hundreds of contacts is a complete waste of time. You need a way to quickly jump to the contacts you frequently call.

That's where the VIP list comes in. Think of it as the Fire phone's speed dial. Adding someone to the list is a snap—on the contact information screen, tap the small white star on the person's picture (or the generic person icon, if there's no picture). The white star turns red. That's all it takes—the person is now on your list. To remove someone, tap the red star and it turns white.

To see the VIP list, swipe or tilt to open the right panel. Tap anyone on the list, and his Contact information appears. Tap the phone icon to place the call.

Mitch Featherston

Mitch Featherston

Call Mobile

Text Mobile

Phone Ringtone
Default ringtone (Olympic)

Answering Calls

WHEN A CALL COMES in on your Fire phone, you'll know in no uncertain terms that someone's trying to reach you. The way you find out when a call is coming in, and the way you answer it, depend on what you're doing when you get the call:

- **If you're doing something else on the Fire phone**, you'll hear a notification, and a screen appears telling you that you've got a call. Tap the green Answer button to answer it. If you don't want to answer the call, tap the red Decline button.

- **If the Fire phone is asleep or locked**, the phone rings and you see information about the caller. Swipe up from the bottom of the screen to answer. Just let it ring if you don't want to answer it.

- **If you're wearing earbuds or listening to music on an external speaker**, the music stops playing and you hear the ringtone in your earbuds or speaker. Answer the call in the usual way. After the call ends, the music starts playing again.

Incoming Call...

Preston Gralla

Author
Work

Decline   Answer

Turning Off the Ring

If the phone starts to ring at an inopportune time, you can turn off the ringtone without turning off the phone or dropping the call. Just press the volume switch at the phone's upper-left side. The ring goes away, but you can still answer the call in any of the usual ways.

Ignoring the Call

Suppose you're just getting to the juicy part of a book you're reading, or you simply don't want to talk to the annoying person calling? You can ignore the call. After five rings, the call goes to voicemail. (It does so even if you've silenced the ring with the volume switch.)

Responding with a Text Message

Don't want to talk, but do want someone to know why you're not answering the phone? Instead of answering the call, you can respond with a text message. When the call comes in, tap the small icon that looks like a badge on the right side of the screen. Tap the text message you want to send—for example, "I'll call you right back." A few minutes after the call ends, the person will get the text.

If you're not happy with the selection of built-in text messages, create your own—"Can't talk; I'm on Mars," for example. Swipe or swivel to pull down the Quick Actions bar and tap Settings. Then tap Phone→"Edit Reply-with-Text-messages" and you'll see the list of messages you have. Tap any, edit the text, and then tap OK.

Conference Calls

HERE'S A FANCY PHONE trick that's great for business and personal use—conference calls. Sitting at a Starbucks, but want people on the other end to think you're in an office with a phone system? The Fire phone lets you conference in multiple people with no extra charges or software (depending on your carrier). And you can also use it for conferencing in friends when you're all trying to decide whether to meet at 6 p.m. before the Red Sox game so you can catch Big Papi at batting practice.

To make a conference call, first make a call as you would normally. Then tap the + Add Call button on the far right. You jump to your contacts list, so you can choose someone to call from there. If you prefer to use the keypad or history list, tap the appropriate icon. If you decide you don't want to make the second call after all, tap the green "Tap to Return to Call" button at the top of the screen.

After you make the second call, the first call is put on hold. To conference in both calls, tap the Merge button that appears onscreen. That's all it takes—you're now on a conference call. The screen shows the names of both callers. Tap the + button to add another caller.

If you prefer to switch back and forth between the two calls, instead tap the On Hold button at the top of the screen.

When you're ready to finish up, tap End Call. If you're in a conference, both calls will be ended. If you're talking to just one of the people on the line, only that call will end, and you'll be automatically switched to the other call.

2:58

Preston Gralla,
Preston Gralla

Conference call

End Call

Call Waiting

THE FIRE PHONE OFFERS call waiting, so it lets you know when you have an incoming call while you're already on a call. You can choose to either answer the new call—and put the first call on hold—or ignore it and let it go to voicemail. If you're a real fast talker, you can even keep both calls going at once and switch back and forth.

When you're on a call and another call comes in, your screen shows you the phone number of the second caller. The phone won't ring or vibrate, and at that point, you're still on the call with your first caller.

TIP If you want, you can turn off call waiting. Go to Settings→Phone→Configure Call Waiting and turn it off on the screen that appears.

You have three choices:

- **Answer: Hold current call**. Answer the call as you would normally, and you can accept the incoming call. When you do this, the first call is put on hold. To switch back and forth between the two calls, tap the On Hold button at the top of the screen. You can also merge the two calls and turn it into a conference call as detailed in the previous section.

- **Answer: End current call.** When you do this, you answer the incoming call and hang up on the one you're already in. Be polite, though, and tell the first caller you need to hang up.

- **Decline.** You can ignore the call by tapping this button (or just letting it ring). Proceed with your first call.

Call Forwarding

CALL FORWARDING LETS YOU have your Fire phone calls rerouted to a different number. That way, when you're at home, you can have your cellphone calls ring on your landline and only have to deal with one phone. Even if the Fire phone is out of commission, you can still forward its calls to another phone and never miss a call.

If you haven't used call forwarding before, first make sure that the service is available from your carrier and activated. Check with the carrier for details.

Once you've done that, turning on call forwarding is a breeze. Go to Settings→Phone→Forward Incoming Calls→Call Forwarding. Turn Call Forwarding on, enter the number where you want your calls sent, and then tap OK.

Caller ID

CALLER ID IS BUILT into the very guts of the Fire phone. That's why you see the phone number, and at times the caller's name, every time you get a call. And when you call someone, she sees your phone number as well.

But if you don't want your number to be displayed that way, you can block it. Go to Settings→Phone→Configure Caller ID→Show My Caller ID. From the screen that appears, you can choose to hide your number, display your number, or default to whatever your cellphone provider normally does.

Visual Voicemail

Visual voicemail is such a great idea, it's surprising it took so long to think of it. It lets you see and manage all your voicemail in a chronological list instead of going through the usual annoying and time-consuming voice prompts. So if you've got 10 messages, you don't have to play them one after another until you find the one you want to hear. Instead, you can just tap to go straight to the message you want.

To set up visual voicemail for the first time, tap the Voicemail button on the keypad (page 59). You'll be connected to an automated system that walks you through setting up a PIN, greeting message, and so on.

Connect to visual voicemail by tapping the voicemail icon, and then enter your PIN. After that you'll see a list of all your calls. Tap any message and then tap the Start button to play it. You can pause by tapping the Pause button and go back through the messages by dragging the scroll bar. To call the person back, tap the Call Back button at the bottom of the screen.

NOTE Make sure to tap to the right of the icon representing the person who made the call, and not the icon itself. If you tap the icon itself, you'll instead see contact information about the caller.

To delete a call after you've tapped it, tap the Trash icon at the bottom of the screen. You can also hold your finger on the call and then select Delete Voicemail. Even if you've deleted the message, though, it lives on. Look at the bottom of the message list, and you'll see an option called Deleted Messages. Tap it to see all the messages you've deleted. From here, you can tap any message to listen to it. To undelete the message, hold your finger on it and select Undelete Voicemail. To kill it permanently, instead select Delete Voicemail.

Changing Your Voicemail Greeting

If you'd like to change the message you recorded when you first set up Visual Voicemail, tap the Voicemail button and then tap the microphone button at the top right of the screen. Tap Custom, then tap the Record button to record a new message. Tap Play to listen to it. If you're not happy with the message, tap Record again and re-record it. When you're satisfied with it, tap the check-mark at the top of the screen. Congratulations—you've just recorded your new greeting.

 # Changing Your Ringtone

NOT HAPPY WITH YOUR Fire phone's built-in ringtone? Go to Settings→ Sounds and Notifications→"Change your ringtone" and tap Sound. On the next screen, tap any ringtone you want to try out, and you'll hear it play. If you want to make that your ringtone, tap the checkmark at the top of the screen. If you want no sound at all, tap None.

Bluetooth Earpieces

BLUETOOTH IS A SHORT-RANGE wireless technology designed to let all kinds of devices connect with one another, exchange or sync files and photos, and let a cellphone serve as a wireless modem for a computer.

With a little work, you can get the Fire phone to do all that. Mostly, though, the phone's Bluetooth capabilities come in handy for hands-free talking with an earpiece. If you've ever seen someone walking down the street, apparently talking to an invisible friend, you've seen Bluetooth in action (unless he really *was* talking to an invisible friend). The small device clips to your ear, and you talk into its microphone and listen in the tiny speaker.

NOTE The earpiece you use for making phone calls is typically monaural, and not designed for listening to music. If you're a music lover, invest in a *stereo* Bluetooth headset.

Pairing with a Bluetooth Earpiece

To use a Bluetooth earpiece with your phone, you'll need to *pair* them—that is, get the two of them talking to each other. The process is a bit geeky, but not difficult. The exact steps may vary a bit depending on the earpiece you're using. But generally, these are the steps you'll take:

On the earpiece, turn on Bluetooth and make it discoverable. In other words, set the earpiece so your Fire phone can find it. Check the earpiece's documentation on how to do so.

On your Fire phone, turn on Bluetooth as well. To do it, pull down the Quick Actions bar and tap the Bluetooth icon so it turns orange. Then tap the Settings icon and select "WiFi& Networks"→"Pair a Bluetooth Device." The phone will scan for any nearby Bluetooth equipment. Your earpiece should show up on the list.

Tap the earpiece's name and type the passcode from the earpiece's manual. The passcode is for security purposes, so no one else can pair with the device. The number is usually between four and six digits, and you'll need to type it within a minute or so. You only need to enter the passcode once. After that, the pairing will happen automatically.

You should now be connected. You still place calls using the phone, but you talk through the earpiece. Check the earpiece's documentation on how to answer calls, control the volume, and so on.

TIP   Having trouble getting your new Bluetooth earpiece to work? Search the Internet for the make, model number, and the word *setup*. If you're having trouble, someone else likely had trouble as well, and you just may find a solution.

Bluetooth and Car Kits

An increasing number of cars include built-in Bluetooth so you can pair your Fire phone with it. This way, you can make calls directly from your car's control panel and hear calls over your car's speakers. You can do other nifty things, such as play music from your phone on your car's entertainment system.

If your car doesn't have built-in Bluetooth, there are plenty of Bluetooth car kits out there. Generally, pairing your phone with a Bluetooth car kit is much the same as pairing it with a Bluetooth earpiece. How you use the car kit varies, of course. In some instances, you can dial a number on the car's touchscreen or answer the phone by pressing a button on the steering wheel.

TIP Don't be lulled into thinking that driving with hands-free calling is safe. Studies show that the danger in talking on a phone while driving isn't related to holding the phone—it's the distraction of holding a conversation while driving.

Text Messaging

WHAT!? YOU SAY YOU use your smartphone to make phone calls? That's *so* early twenty-first century! Fittingly, the Fire phone lets you send and receive text messages—and lets you add pictures and videos to them as well.

When you send text messages, you use the SMS (Short Message Service), which limits you to 160 characters (including spaces and punctuation), which comes out to a sentence or two. That may sound short, but in a world where Twitter limits you to messages of 140 characters, 160 characters can suddenly seem like a lot of space.

Text messaging doesn't come free. You'll have to pay extra, either for a monthly plan or for individual text messages. Check with your wireless provider for details.

NOTE The charges for text messaging are for messages you *receive* as well as those you send.

Videochat on the Fire

Videochat is a more intimate mode of communication than text messages or even voice calls. A videochat app establishes a direct connection between you and another person or people. The Fire phone doesn't have its own built-in videochat app, but there are several free ones that you can download. To find one, search for *video chat* in the Amazon Appstore. Here are a couple of cool options:

Tango lets you videochat with other people who use Tango on Android phones, iOS devices, Windows Phones, and Windows computers. You can switch back and forth between video and audio and audio only, and you can text with it as well as share files.

Another good free one is oovoo, which lets you have videochats with up to 12 other people. You can also instant message with it, and record your videochats and upload them to YouTube. As with Tango, you can make audio-only calls using the service as well. It works on Android phones, iOS devices, Windows Phones, PCs, and Macs.

Unfortunately, at this writing, the Fire phone doesn't let you use two of the most popular videochat programs—Skype and Google Hangouts.

Receiving a Text Message

When you get a text message, the Fire phone plays a notification sound. What happens next depends upon whether the phone is active or asleep:

- **If you're using the phone**, you hear a notification sound, and a message appears across the top of your screen. Tap it.

- **If the phone is asleep**, you hear a notification sound. Wake up the phone, and then tap the Messaging icon on the Apps Grid.

In either case, you go straight to a list of your most recent text messages, those you've sent as well as those that have been sent to you. Depending on the length of the message, you'll see the whole message (if it's short) or only part of it (if it's long).

TIP When someone sends you a text message with links in it, the links are live. Tap a web address to visit it in your browser, or tap a phone number to dial the number for a voice call.

To read the message you were just sent, tap it. You see the message in a text balloon, and if it's part of an ongoing conversation of messages, you see each message.

Text messagers are encountering an unpleasant fact of texting life—spam. It's not nearly as prevalent as email spam, but you'll most likely get some at some point in your messaging life.

To respond, tap in your message using the keyboard, and then tap the Send icon next to your text message. Off your message goes, instantly. You see the record of your message appear in a text balloon. If your friend texts back, you see it in a text balloon...and so on.

Sending a Text Message

To send a text message if you haven't received one, tap the Messaging icon. The list of all your messages appears. Tap the Compose icon at upper right, and a screen appears where you create your text message. There are several ways to tell the Fire phone where to send your text message, both accessible from the To field:

- **Type a name into the field**. The phone looks through your contacts and displays any matches. Tap the contact to whose cellphone you want to send a text message.

- **Tap a phone number into the field**.

- **Tap the icon of a person** next to the recipient field, and you're sent to your Contacts list, where you can choose a recipient.

After that, type your message in the message field and tap the Send icon, and your message goes on its merry way.

Adding Pictures and Video

On the Fire phone, the term *text message* is an understatement. It's a breeze to send a photo or video by SMS. You can even take a photo or record a video and embed that as well. When you're composing your text message, tap the icon of a camera next to the text box and select "Capture a Photo," "Capture a Video," or "Choose Existing." If you choose to capture a photo or video, you'll be sent to the phone's Camera app. Take the video or photo as you would normally (page 116), and it gets embedded in the message itself—you'll see it there before you send it.

If you select Choose Existing, you're sent to the Photos app (which also includes videos). Find the picture you want, tap the checkmark, and the item gets embedded in the message.

> **NOTE** If the photo or video is too large to send, the Fire phone will compress it for you. It will take a few seconds to compress, depending on its size.

When you're viewing a contact, you can send a text message right from that screen. Tap the Text Mobile icon next to a phone number to address a text message to that number. Anywhere you view your contacts, you're only a tap or two away from sending a text message. Similarly, viewing a picture or a video, tap it so the top menu appears, and then tap the Share button. From the screen that appears, tap the Messaging icon. The photo will be embedded into a text message. Type any text you want, and then add the recipient and send it as you would any other text message.

Controlling the Fire with Your Voice

YOU CAN USE YOUR voice to do more than just talk on the phone. You can also talk to the Fire phone and have it do rudimentary tasks for you, including making phone calls, sending text messages, sending emails, and searching the Web.

NOTE Don't bother saying, "OK, Google." The Fire phone's voice controls are built by Amazon and aren't the same as the Google search feature found on many other Android phones. You can't do nearly as much by voice as you can with "OK, Google"—you can't, for example, get directions or maps using the Fire's mapping app, launch apps, or play music.

To control the Fire with your voice, press and hold the Home key. A microphone appears onscreen, along with the word "Hello." You can then issue any one of four commands by talking into the phone:

Making a Phone Call

After you hold down the Home button, say, "Make a call." After you're prompted, say the name of the person you want the Fire phone to call. If there's only one number for that person, the phone will immediately call it.

If the person has more than one number, for example, Home and Mobile, the Fire phone will show both numbers and ask which you want to call. Tell it which to call, and the phone calls for you. If you're the impatient sort, you can simplify the whole thing and simply say the name of the person and which of her phones you want to call—for example, "Call Lydia Gralla's mobile phone."

Sending a Text Message

You send a text message in much the same way you make a phone call: Hold down the Home button; say, "Send a text message"; when prompted say the name of the person to whom you want to send a text; and when prompted again, dictate your message. As with making a phone call, if the person has more than one phone number to which a text can be sent, you'll be asked which one to send it to.

After you dictate the message, you'll see the text of the message and see three buttons onscreen: Send, Cancel, and Edit. Say or tap Send to send it on its way. Say or tap Cancel if you decide you don't want to send it.

You can also edit the message. Tap the Edit button, and you'll instead get the usual text message screen (page 79). From there, you can edit your message and handle it as you would any text message.

WARNING Be careful to tap the Edit button, and not say "Edit." If you say "Edit," the message will be erased and you'll have to start over and dictate your message again.

You can take a shortcut here, as you can with making a phone call, doing every-thing in one fell swoop by saying, for example, "Tell Lydia Gralla, How about dinner tonight?"

Sending an Email

To send an email, hold down the Home button and say "Send an email." When prompted, say your intended recipient's name, and when prompted again, dictate the subject of your email, and then the message itself. As with making a phone call, if the person has more than one email address, you'll be asked which one to send it to. When you're done, you'll see the text of the email and three buttons: Send, Cancel, and Edit. Say or tap Send to send it on its way or Cancel to back out. Say or tap Edit to edit it, but be prepared for a long process, because telling the Fire what changes to make can take a long time, particularly if it doesn't recognize your words correctly. A better bet is to tap the Edit but-ton. You can then edit and send the message using the Fire phone's usual email program.

Searching the Web

To search the Web, hold down the Home button, and then say, "Search" or "Search the Web." Dictate your search term. The Fire launches the Silk browser, searches the Web, and displays the results as it does for any web search (page 174).

Media

You'll learn to:

- Use the Kindle app to read books
- Use notes, highlighting, and bookmarks
- Buy books from the Amazon bookstore
- Subscribe to magazines

CHAPTER 4

Books and Magazines

THE FIRE PHONE WEARS its Amazon heritage proudly. It's one of the reasons many people buy it. Not surprisingly, some of its most important features are tied directly to Amazon. You can read and manage your books with the same Kindle reader you use on other devices.

Using the Kindle App

THE FIRE PHONE MAY not be a tablet, but its 4.7-inch 1280 x 720 resolution screen does an impressive job with books. No, it's not well suited for big picture books, but for most everything else, it delivers what you need. On the Carousel or Apps Grid, tap the Kindle icon, and it springs into action.

TIP You can also access your Kindle library by swiping from the left of the Home screen (or using the one-handed Tilt gesture) to access the left panel and tapping Books, which opens the Kindle app.

You'll see any Kindle books you own already there waiting for you. To read or download a book, tap its thumbnail. If the book isn't downloaded yet, it will take a short while for the book to download to the phone.

When you download a book to the phone for the first time, you'll see an orange New sash in the thumbnail's upper-right corner. Any book you've already downloaded will have an orange badge on it showing how much of the book you've read.

Turning Pages and Navigating a Book

Where you start when you open a book depends on whether you've already been reading it in the Kindle app or are just opening it for the first time. If it's the first time, you're sent to the beginning of the book. If you've already been reading it, you head to where you left off.

> **NOTE** The Fire's Kindle app syncs with all other Kindle devices and readers. So if you've started reading a book on a Kindle Fire HD, for example, when you continue reading it on the Fire phone, you get sent right to the page you left off when you were reading it on the Fire HD.

You use the same gestures to turn the pages in the Kindle reader that you do everywhere else on the Fire phone. Swipe to the left or right to page ahead or back through the book. You can also tap anywhere on the right edge to move forward a page and anywhere on the left edge to move back a page.

NOTE Someday you'll be able to use the Dynamic Perspective Auto-Scroll feature (page 28) to move through a book by tilting the phone away from or toward you. It wasn't included in the initial Fire phone release, but Amazon is working on it. Once Auto-Scroll is added to the app, you'll be able to customize it to scroll at the pace you set without your having to keep tilting.

For faster navigation, you can make the Location bar appear at the bottom of the page by tapping anywhere in the middle of the screen. Down at the bottom, you'll see a slider that shows you your location in the book and the total number of pages. You also get to see what percentage of the book you've finished reading as of your current location. To move forward or back, drag the slider to where you want to be.

Navigating with the left panel

You have one more way to get around a book—using the Kindle reader's left panel. Swipe from the left or use the one-handed Tilt gesture (page 26) to reveal it. Here are all the ways you can navigate from there:

- **Search Inside This Book.** Tap here, and a search box appears. Type the word or name you're looking for, tap Done, and you'll see all the places in the book where it appears. You see the term in context, with the surrounding sentence or two. Tap any result to head to that page.

- **Go to Page or Location.** Tap here, and you can jump to a page or location by typing the page or location number. In Kindle books, *location* is a more precise way to navigate than by page number. Every location is a specific spot in the book, regardless of how it's paginated.

- **Sync to Furthest Page Read.** If you've been jumping around in a book, either on the Fire phone or on a different device in addition to your Fire, just tap here, and you'll jump straight to the furthest page you've read on any of your devices.

- **Table of Contents.** Tap any chapter title to head there.

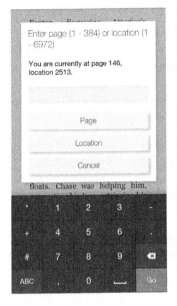

Changing Your Kindle Options

The Location bar does much more than just let you hop around a book. Tap in the middle of any page to activate it, and here's what you can do:

- **Change font and size** Aa. Tap here, and you'll be able to change the font, font size, text and background colors, and spacing between lines.

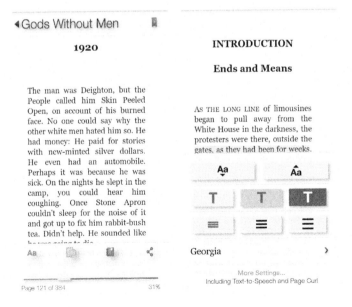

- **X-Ray** 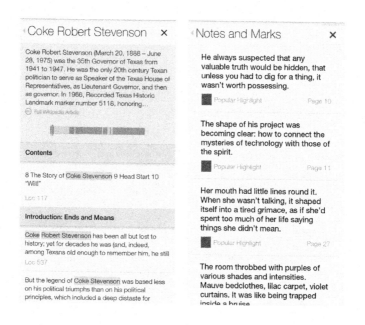. This unique-to-Kindle feature is as if someone pre-read the book and marked where and how often important characters and terms are mentioned. If you forget who someone is or need more information about a term, tap X-Ray. Then tap any of the listed characters or words, and you'll come to a screen that gives a brief biography of the person, or more information about the term. There's also a list of everywhere the person or term is mentioned. Tap an item on the list to go to the page.

You'll notice a gray bar associated with each character and term, with light- and dark-blue bands in it. The gray bar is a visual representation of the entire book. The light-blue bands show you where the character or term appears on the page you're currently displaying. The darker bands show where it appears relative to the whole book.

To return to the book when you're done, tap the X at the upper right (labeled Close if you peek; see page 23) or the X-Ray button at the upper left.

NOTE   The X-Ray feature isn't available for every book. If it's not available for the book you're reading, the X-Ray button is grayed out.

- **Notes and Marks** . Tap here to see notes and highlights you've created yourself (page 96) as well as passages that other Kindle users have highlighted.

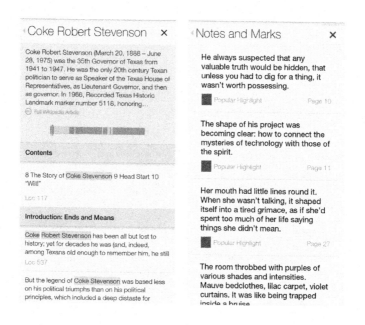

- **Share** ⋖. Tap here to share the book with others via social networking apps, email, Bluetooth, and other ways. You'll share the name of the book as well as a link to its page on Amazon.

- **Narration** ▶. If there's any audio narration available for the book, a narration button appears at the very bottom of the screen. Tap it to have the book read to you. This narration will be robotic-sounding; it isn't the actual audiobook, read by the author or a professional narrator. For that, you have to pay. To check whether professional narration is available for a book, when you have the book open in the Kindle app, open the left panel and scroll down. If you see Add Professional Narration, then narration is available. Tap that link for details on how to purchase it.

- **Bookmark** ▯. Tap the little bookmark icon at the top right, and then tap Add Bookmark to make the page easy to return to later. After you add bookmarks, they all appear when you tap the Bookmark button so you can jump right to any of them. You can also delete any of your bookmarks by tapping the bookmark and then tapping Remove Bookmark.

NOTE When you're on a bookmarked page, you'll see a blue triangle in the top-right corner.

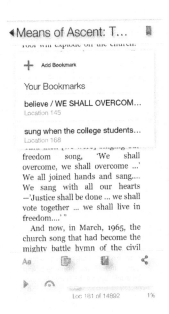

◀ Means of Ascent: T...

Your will explode on the church.

➕ Add Bookmark

Your Bookmarks

believe / WE SHALL OVERCOM...
Location 145

sung when the college students...
Location 168

freedom song, 'We shall overcome, we shall overcome ...' We all joined hands and sang.... We sang with all our hearts —'Justice shall be done ... we shall vote together ... we shall live in freedom....' "

And now, in March, 1965, the church song that had become the mighty battle hymn of the civil

Aa

Loc. 181 of 14892   1%

the White House gates—with the pickets singing "We Shall Overcome" as if to tell him to his face, If you won't help us, we'll win without you. But one of the assistants riding with him had worked for him for almost twenty years, and saw his expression, and knew what it meant. "He heard," Horace Busby recalls.

WITH ALMOST the first words of his speech, the audience—the congressmen and Senators with whom he had served, the Cabinet members he had appointed, the black-robed Justices of the Supreme Court, the Ambassadors of other nations, a few in robes of far-off countries as if to dramatize that the world as well as America was listening, the packed galleries rimming him above—knew that Lyndon Johnson intended to take the cause of civil rights further

27 mins left in chapter   2%

Notes and Highlighting

If you like to highlight memorable sentences and write notes in the margin when you read a print book, you'll be glad to know the Fire phone makes it easy to add notes or highlighting to any Kindle book:

- **Hold your finger on a word or passage you want to highlight or create a note for**. A magnifying rectangle appears around the word or passage. You can expand to include more words by dragging the positioning triangles at either end.

- **Choose an action from the pop-up menu that appears.** Above the high-lighting window, you see your choices:

  - Tap any of the colors to add that highlight to the text.

  - Tap the Share icon to spread the words via social networking, email, and more.

  - Tap the Note icon (pencil and piece of paper) to write a note.

  - Tap the Menu icon (three vertical dots) for two options: "Search in Book," and "Search the Web."

NOTE The Menu button ⋮ shows up in many places on the Fire phone. Tapping it usually reveals a few contextual commands.

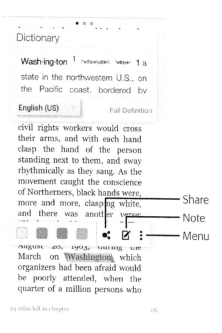

When you hold your finger on some text, the top or bottom of the screen displays information about it—for example, a dictionary definition, or information from Wikipedia. If it's a foreign word, Bing Translator will show a translation. Only one option (for example, Wikipedia information) will appear at the top or bottom of the screen. But if you swipe that information window to the left or right, you'll see more options, including the dictionary definition and a translation, if available.

- **Review your notes and highlights**. Tap the middle of the screen to display the Location bar, and then tap the Notes icon to see your notes. In addition to displaying your notes, it displays passages that other readers have highlighted.

When you look through highlights in a book, you'll see not just your highlights, but passages that other readers of the Kindle book have highlighted as well. Not every highlight appears, only those that Amazon has determined are popular with many readers.

One of the Fire phone's biggest benefits is that it gives you a free year of the Amazon Prime service, which normally costs $99 a year. That's a pretty good deal, and one that no other phone can offer. (See page 32 for the full story on Amazon Prime.)

Buying Books

NOT BY ACCIDENT, IT'S just as easy to buy books on your Fire phone as it is to read them. In the Kindle app, tap the shopping cart icon at top right (labeled "Store" when peeking) and you'll get to the bookstore. The bookstore shows books that it believes you might want to buy based on your past buying habits, as well as a variety of categories, such as daily deals, best sellers, and so on. Scroll down to see them all.

To get more details about any book, tap it. You'll find reviews, price, and more, usually including a way to download a sample. And, of course, you can always buy the book here as well.

It's always worthwhile to check out reader reviews and ratings. Note that the star rating is an average of all ratings, and right next to it is the total number of people who have rated the book. This number is more important than you think. If only a relatively small number of people have rated a book—for example just 20 or so—that rating may have come from family and friends willing to write good reviews with good ratings. But when there are hundreds of ratings, the average is far more trustworthy.

Tap the rating to see individual reviews and ratings. Don't expect a uniformity of opinion, because taste in books is subjective. Instead, look for the overall drift of the reviews.

As you're browsing through the bookstore, you can use the Fire phone's Peek feature to reveal information about each book right on the thumbnail, including the price and reviewer ratings. Just tilt the phone slightly to the left or right.

> **NOTE** If you subscribe to Amazon Prime, you get access to a lending library, and to new books before they're published. For details, see page 35.

To browse categories, open the left panel, and then select the category of books you want to browse. To find more recommended titles or best sellers, open the right panel and tap to choose from the list.

NOTE You'll generally find Amazon's recommendations useful, although you're un-likely to come across any unusual finds. Often the recommendations are books written by authors whose other books you've previously bought. You'll also find books in the same genres in which you've already purchased books. They're basically the same as what you get on the recommendations tab on Amazon's home page.

Getting Books from the Library and the Web

BUYING BOOKS FROM AMAZON—or getting them free (see page 35)—isn't your only choice when it comes to reading books on your Fire phone. You can also check out Kindle books from your local library or download them from the Web.

NOTE The details of the borrowing procedure may vary from library to library. This section describes how it's most commonly done. For more information, check your local library.

First, log into your library's website using your user name and password. Then search for a book as you would normally. In some instances, you'll be able to specifically say you're looking for an ebook, but in other instances, you'll do a search and see both ebooks and print books. Your library may also offer audio-books. If so, you can borrow them the same way as Kindle books and listen to them on your Fire phone.

Click the book you're interested in. If it's available for lending, you'll see that somewhere on the page. Theoretically, the library could make as many copies as it wants available for checking out electronically, because the ebook isn't a physical thing that gets checked out (even when it's checked out, it's still on the library's servers). However, publishers limit the number of copies of the book that can be checked out at once, because they're worried that otherwise, many fewer people will buy the book. You're likely to encounter a waiting list for popular titles...not unlike the waiting list you might find at your library for the print version of new bestsellers.

Once you find an available book you want to download, you'll typically find it in multiple formats. Select Kindle Book, and then confirm to download it.

You'll next be sent to the Amazon website, where you're probably already signed in. If not, sign in using the account you use on your Fire phone. (If you don't do that, you won't be able to get the book onto your phone.) Click Get Library Book. The book will be sent to your Fire via WiFi automatically using an Amazon feature called WhisperSync.

Bookshelf

Note: If you download a title, its Return Title button will disappear. Learn about your other return options here.

If a title is available to renew, the option will become available within 3 days of its expiration.

Howards End
E. M. Forster

The Final Storm
Jeff Shaara

Howards End [Kindle Edition]
E. M. Forster (Author)

- Digital library books require an active Wi-Fi connection for wireless delivery to a Kindle device. Library books will not be delivered via your Kindle's 3G connection.
- Kindle automatically backs up all of your notes and highlights. If you choose to purchase the book at a later date from the Kindle Store or check it out again, all your notes and highlights will be preserved.

Expiration Date: September 21, 2014

You may also end your Library Loan early by visiting Manage Your Content and Devices and choosing "Return This Book" or return the book now by clicking below.

Return Book

- Length: 324 pages (Contains Real Page Numbers)
- Don't have a Kindle? Get your Kindle here.

You are signed in as
preston@grafla.com
Use a different Amazon account

Get library book

Deliver to:
Preston's Fire Phone

Buy Now with 1-Click

Free
Kindle
Reading
App

Anybody can read Kindle books—even without a Kindle device—with the FREE Kindle app for smartphones and tablets.

> **NOTE** You may also want to try borrowing library books using the free OverDrive software. Some libraries, in fact, require it. You can use the OverDrive software not just for Kindle books, but also for books in other formats, including PDFs. And you can also use the OverDrive software for downloading books from websites. For details on how to use it, go to *www.overdrive.com*.

Reading Magazines

YOUR FIRE PHONE LETS you read magazines and newspapers, not just books. To get started, tap the Newsstand icon on the Apps Grid. (You can also open the left panel and tap Newsstand.) If you don't yet have any magazine or newspaper subscriptions, you'll see a message that your library is empty. To search for a magazine to buy, tap the shopping cart icon (labeled "Store" when you're peeking) and type a search term. You can search both by title and keyword. For example, if you search for the keyword *news*, you'll find many matches, including the *San Jose Mercury News* and *Time* magazine. If you prefer to browse rather than search, simply go through the categories, including "First 30 Days Free," "Fall Family Meals," "Food and Recipes," "Crafts & Hobbies," and many others.

Tap any magazine to get more information, including the price of an annual subscription or single issue. Tap the appropriate button to subscribe or buy. As with Kindle books, you can browse magazines by category (swipe or tilt to reveal the left panel) or recommended titles (right panel). To find a magazine in your own library of subscriptions, tap the Search button.

> **NOTE** You can use the Peek feature to see more information about magazines on their thumbnails, in the same way as in the Amazon bookstore. Tilt the phone to the right to reveal the details.

When you read a magazine, you won't get the full suite of features and tools you do when reading books—bookmarking is about all you get. However, you can display the magazine's table of contents by displaying the left panel. You can also magnify any text by holding your finger over it.

If you decide not to buy something in the Newsstand, tab the Library icon (nine squares in a square grid, labeled "Library" when peeking) to return to your Library.

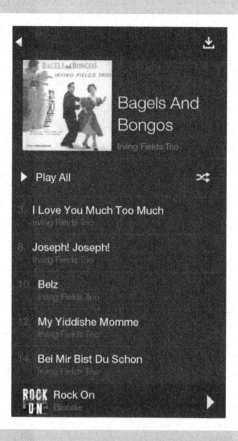

Bagels And Bongos
Irving Fields Trio

▶ Play All

3. I Love You Much Too Much
Irving Fields Trio

8. Joseph! Joseph!
Irving Fields Trio

10. Belz
Irving Fields Trio

12. My Yiddishe Momme
Irving Fields Trio

14. Bei Mir Bist Du Schon
Irving Fields Trio

ROCK ON Rock On
Blondie

You'll learn to:
- View photos and videos in the Gallery

- Take photos and videos

- Watch videos, including free videos from Amazon Prime

- Listen to music, including free music from Amazon Prime

CHAPTER 5

Photos, Video, and Music

YOUR FIRE PHONE IS a multimedia and entertainment marvel. Like all smart-phones, it has a camera so you can shoot your own photos and videos, but as always, there's a special Fire twist—moving images called *lenticular* photos. You can edit your photos right on the Fire phone as well.

Furthermore, you can enjoy photos, videos, and music from a variety of sources: transferred from a PC or Mac (page 141), received as email attachments, or streamed to your phone. You can also watch movies and TV shows from Amazon Prime Instant Video.

Using the Photos App

YOU VIEW PHOTOS ON your Fire phone in the Photos app. Tap its icon either on the Carousel or the Apps Grid. The first time you open it, you'll see a screen with a grid of square thumbnails of all your photos and videos, four across. Turn your Fire phone sideways, and instead of a regular, even grid, your thumbnails will display according to the orientation they were taken—horizontal pictures horizontally, vertical pictures vertically. The word "All" appears at upper left, confirming that you're seeing all your photos. (If your phone doesn't reorient the screen when you turn it, you need to enable screen rotation in Settings; see page 306.)

Subsequent times you launch the Photos app, you'll go right back to the last thing you were doing before you left—viewing a single photo, for example.

Tap a photo to view it. A variety of buttons appear onscreen for a moment or two and then disappear. Tap the photo to make the buttons reappear.

Depending on the photo's length-to-width ratio, it may not fill the entire screen. If it doesn't, you'll see black space along the sides or at the top and bottom.

Up at the top left of the screen you'll see numbers, such as "42 of 434." That tells you how many photos you've got on your phone (434, in this case), and where the current photo is compared to the rest of your photos (the 42nd photo). Tap the left-facing arrow to go back to browsing through all your photos.

Don't forget to use the one-handed Peek gesture when you're browsing your photos and videos. Tilt the phone slightly to the right or left. Next to each you'll see information about it—the length of the video, or the date the photo was taken.

Look down at the bottom when you're in single-photo view. You'll see four icons that do the following, from left to right:

- **Camera** launches the camera. For more details about how to use it, turn to page 116.

- **Share** lets you share your photos in many different ways: email, text messaging, sending to Bluetooth-enabled devices (page 74), via social media, and more. The exact ways you can share will vary according to what apps you've got on your Fire.

- **Trash** deletes the photo.

- **Edit** launches the Photo Editor, which gives you a solid set of tools for editing photos, including cropping, rotating, changing colors, and more. You can also add some cool effects—see the box on page 110.

Viewing Pictures

Now that you've got photos on your screen, the fun begins—viewing them in different ways and flicking through them:

- **Zooming** means magnifying a photo, and the power is at your fingertips—literally. Double-tap any part of the photo, and you zoom in on that area; double-tap again and you zoom out. You can also use the thumb-and-forefinger spread technique to zoom in more precise increments. Once you've zoomed in this way, you can zoom back out by using the two-finger pinch technique. (Flip back to page 20 for a refresher on all these techniques.)

- **Panning** means to move the photo around the screen after you've zoomed in, so you can see different areas. Use your finger to drag the photo around. As with Zooming, panning works the same whether you're holding the phone horizontally or vertically.

- **Rotating** means to turn your phone 90 degrees so it's sideways. When you do so, the photo rotates and fills the screen using the new orientation. This technique is especially useful when you have a horizontal photo that looks small when your Fire phone is in its normal vertical position. Rotate the phone, and like magic, the picture rotates and fills the screen properly. Similarly, if you're viewing a vertical picture while holding the phone horizontally, simply rotate the phone 90 degrees, and your photo rotates as well.

- **Flicking** advances you to the next or previous photo in your list. Flick from right to left to view the next photo, and from left to right to view the previous one.

Edit Away

The Fire phone gives you surprisingly powerful photo-editing tools. True, they're not up to the level of a high-powered, expensive piece of software like Photoshop, but you can't beat free.

For a start, there's the Enhance feature. Just tap the icon at far left, and Enhance does its magic, adjusting the brightness, colors, and so on. If you're looking for a no-muss, no-fuss way to improve your photos, it's a great place to start.

On the editor's main screen there are plenty of other tools as well. You can crop, rotate, and resize any photo. The Orientation tool is surprisingly robust. Tap it, and you can rotate the photo to the right or left, as well as flip it horizontally or vertically.

Make sure that you flick through the icons at the bottom of the screen, because there are plenty of tools hidden off to the left. You can remove red eye, add special effects, annotate your photo with text, and more. You can even add decorations like stickers and frames.

Working with Multiple Photos

IN THE PREVIOUS SECTION, you learned about ways to work with individual photos. But you're not limited to slogging along one photo at a time: the Fire phone lets you work with multiple photos at once—delete them, share them, and so on.

When you're viewing photos in the gallery view, tap the Select button at the upper right of the screen—it looks like a checkmark. The words "Select items" appear. Tap the photos you want to select to activate their checkboxes. At the top of the screen, you see how many photos you've selected.

Now you can perform actions on the photos you've selected. At the bottom of the screen are the buttons for sharing them and deleting them, exactly as for individual photos (page 108). If you change your mind and don't want to do anything with these photos, tap the X (labeled "Cancel" when peeking) at the top right, and the checkmarks go away.

The Left and Right Panels

The left and right panels provide some useful tools you might otherwise not see. Wherever you are in the Photos app—whether viewing a group of photos or viewing an individual photo—when you open the left panel, you'll see different ways to filter which photos you see. So, for example, you can choose to see only videos or only pictures you took on the Fire phone's camera (the Camera Roll option), and so on.

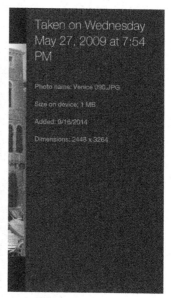

The right panel works only when you're viewing an individual photo. Open it, and you'll get details about the photo you're looking at, including when the photo was taken, its file name, its size, the date it was added to your Fire phone, and the photo's dimensions.

NOTE To open the left panel, swipe from the left of the screen or tilt the phone to the right. To open the right panel, do the opposite. See page 22 for the full scoop on Fire phone gestures.

Videos in the Gallery

THE WAY YOU VIEW and work with videos in the Fire is essentially identical to the way you work with photos, with a few minor differences:

- You can identify videos in the gallery by the right-facing triangles on their thumbnails.

- Generally, you should rotate your Fire phone by 90 degrees when you're viewing videos, since they're usually taken with a horizontal orientation.

- When you tap a video, it opens with a right-facing triangle—the Play button. Before you play the video, buttons appear at the bottom of the screen to let you launch the camera, share the video with others, and delete it.

- Tap the triangle to play the video. When the video starts, a white bar moves along a progress bar to show where you are in the video. Drag the white bar forward or back to move forward or back in the video. Tap the Pause button to pause the video; tap the Play button to start playing it again. You'll also see the total time of the video, and for how long the current video has played.

Photos and Videos in the Cloud

WITH YOUR FIRE PHONE and other Amazon devices such as the Kindle Fire, you also get an Amazon Cloud account, and some photos and videos from your device automatically sync to your cloud account. For example, photos and videos you take with the Fire phone's camera, and photos and videos you get via email, are automatically copied to the cloud. However, photos and videos that you copy from your PC or Mac to the Pictures folder on your Fire phone (see page 144 for details) are *not* copied to the cloud.

Viewing Your Cloud Drive on the Fire Phone

The cloud account has photos and videos from your other Amazon devices as well. And if you want to view and download those to your Fire phone, it's simple to do. Display the left panel and tap Cloud Drive. The screen shows you all the albums containing photos and videos you have in Amazon Cloud. Tap any album to browse through it. Then press and hold any photo or video and then select Download to get it on your phone. You also get options for deleting and sharing the file.

If you press and hold a video or photo in the cloud that's already on your Fire phone (such as videos or photos you've taken yourself), then instead of seeing a Download option, you'll get two different options: Delete and Share (which gives you the same options for sharing as it does for photos already on your Fire phone; see page 108). And if you press and hold your finger on the photo that appears just above a folder of photos—for example, the Your Cloud Drive or Fire Camera Roll folder—you'll see two options: "Pin to Home Grid" and "Download." You won't see the Delete or Share options.

Accessing Your Cloud Drive on the Web

Photos and files stored in the Amazon Cloud are also available on the Web, so you can see them in any web browser. Go to *www.amazon.com* and choose Your Account→Your Cloud Drive. (Or go directly to *https://www.amazon.com/cloud-drive* and sign into your Amazon account.) You'll see a list of all your files. Click any, and you can perform a variety of actions on them, including downloading, sharing, copying, deleting, moving (to a different folder), and renaming.

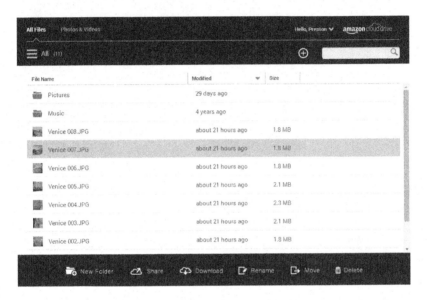

Click Photos & Videos at the top of the screen, and you'll see just your photos and videos, arranged as rows of thumbnails, organized by date. Click any to see it in its own window. When it's in its own window, mouse over it to see tools for downloading and deleting it.

NOTE The organization of your photos and videos in the Amazon Cloud on the Web is different from what you see when you view your Cloud Drive in the left panel of the Photos app on your Fire phone. On the Fire, you'll see folders, including Fire Camera Roll, File Attachments, and Your Cloud Drive (these are photos that you upload to the Amazon Cloud Drive on the Web). However, the web interface doesn't organize photos into those folders. Instead, it organizes all photos by the date they were taken.

Taking Still Photos

THE FIRE PHONE IS not just great for viewing photos, but for taking them as well. It sports a 13-megapixel camera that takes sharp, vivid photos. And it also has a front-facing 2.1-megapixel camera that's ideal for selfies and videochats (see the box on page 77).

Using the camera is point-and-shoot simple:

1. Tap the camera icon on the Carousel or Apps Grid, or press the camera button—it's about a third of the way down on the phone's left side (page 56). Make sure not to press and hold, because if you do, you'll launch Firefly (page 12).

2. Frame your shot on the screen. You can use the camera in the normal verti-cal orientation, or turn it 90 degrees for a wider shot. Zoom in by spread-ing your fingers apart on the screen and zoom back out by pinching them together.

3. Tap the shutter button. It's at the bottom of the screen if you're holding the phone vertically or on the right side if you're holding it horizontally.

You'll hear the familiar snapping sound of a photo being taken. A small thumb-nail of your new photo is displayed in the lower-left corner of the viewfinder if you're taking a picture in portrait mode, and the lower right if you're taking it in landscape mode. Tap it to view the big image. You can zoom in and out in the image by using the normal zoom in and zoom out finger motions.

NOTE You can also transfer photos onto your Fire by copying them from your computer. See page 141.

Using the Onscreen Controls

THE CAMERA HAS A variety of convenient onscreen controls.

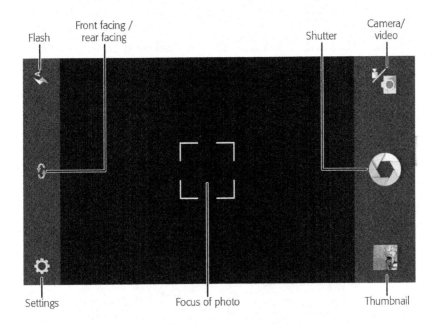

Flash

Front facing / rear facing

Shutter

Camera/ video

Settings

Focus of photo

Thumbnail

- **Flash** turns the flash on or off. Or you can select Auto and let the Fire phone decide when to use its flash.

- **Front-facing/rear facing** switches between the front-facing camera (the one pointed at you) and the normal, rear-facing camera. Use the camera facing you for a self-portrait (think of it as the "selfie" button) or during videochats.

- **Settings** lets you choose from several settings. HDR (high dynamic range) is best for when there are both bright spots and dark shadows in the same scene—for example, when someone stands in front of a window on a bright day. In Image Review, you'll see a preview of the photo right after you take it. That way, you can delete it if you know you don't want to keep it. You can also choose Lenticular and Panorama modes.

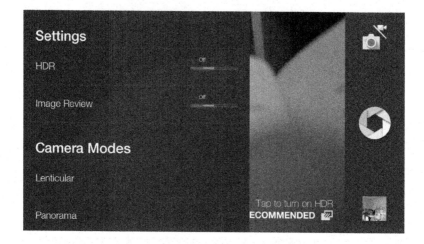

- In **Lenticular** mode, you take a series of images (up to 11) and the Photos app then stitches them together to create a single "moving" image, like a fast video or a 360-degree video around an unmoving object. You see this moving image by tilting the phone when viewing the photos.

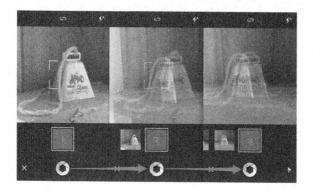

Lenticular Photos: Not Quite Dynamic Perspective

When you take a lenticular photo, you see the moving image using the Fire phone's Dynamic Perspective feature. So you might think that you can use that image as a Dynamic Perspective Lock screen. Unfortunately, you can't. If you choose the lenticular photo to be your Lock screen image, the phone will merely use the first image in the lenticular sequence. So you won't be able to see the same image from different perspectives, as you can with Dynamic Perspective Lock screens. In addition, when you take photos in lenticular mode, they're square in shape, even though the Fire phone usually takes and displays rectangular photos.

In addition, when you share lenticular photos with others, the Photos app converts them to animated .gif files.

- In **Panorama** mode, the app stitches together multiple images to create a single horizontal or vertical panorama. When you choose this mode, the app prompts you to move the camera in a direction indicated by an arrow. When you're done, the app creates the panorama for you.

- **Camera/Video** lets you choose either the still camera or the video camera.

- **Shutter** snaps the picture. When you're in video mode, instead of seeing an icon of a shutter, you'll see a white circle with a small red dot in the middle of it.

- **Thumbnail** shows a thumbnail of the previous photo you've taken. Tap it to see the photo in the Photos app.

Taking Video

TO TAKE A VIDEO, first tap the Camera/Video button and then tap the white circle with a small red dot in the middle of it. The video immediately starts recording, and the white circle turns red. At the upper left, you see how many minutes and seconds of video you've shot and a blinking red button that indicates you're recording video.

When you're recording video, a big red circle appears at the right side of the screen if you're holding the phone horizontally (and you should be holding it horizontally; see the note below). Tap the red circle to stop recording. You can zoom in and out in the same way that you can with the still camera, by pinching or spreading your fingers or using the volume buttons. You can also take a still photo at the same time you're recording video—just tap the shutter button at the upper right.

 NOTE Do you, a friend, or a loved one suffer from "vertical video syndrome"? This fictitious disease afflicts people who hold their phones vertically when taking videos, which means that they will look small and squashed when played the way videos are supposed to be played—in horizontal orientation, as when on a television or theater screen. In June 2012, a tongue-in-cheek video called "Vertical Video Syndrome—a PSA" was uploaded to YouTube warning about this threat. It has over 5 million views.

Watching Video with Amazon Prime Instant Video

AMAZON PRIME INCLUDES FREE video streaming and video downloads. When you buy your Fire phone, you get a free year of Prime (page 32). That means plenty of video watching.

To watch or download Instant Videos, tap the Instant Video icon on the Carousel or Apps Grid to launch the app. To find a video you want to watch, browse by scrolling down to find a category such as Recommended Movies, Recommended TV Shows, and so on. You can use the Peek feature (page 23) to see ratings for each video.

When you find a video you want to watch, tap it. If the video is available for free as part of your Prime subscription, the next screen tells you so. Otherwise, you'll see how much it costs to rent it or buy it. If you want, scroll down to see more information about the video, including information from the IMDb video site.

To play the video, tap the Watch bar below the video's rating. You can also add the video to your Watchlist and return to it later (page 126), or watch a trailer. If it's a TV show, you can also watch other episodes of the show. Scroll down to the episode you want to watch. Tap the arrow to watch the episode.

TIP If you prefer, you can download the video and store it on your Fire phone instead of streaming it. You may want to do this if you know you're going to be away from a WiFi network, or if you know you want to watch the video multiple times. See page 125.

The video will play from beginning to end. Sit back and enjoy the show. If you want to pause, go forward or back, jump to the next or previous scene, change the volume, and so on, tap the screen to reveal the controls. Tap again to make them disappear.

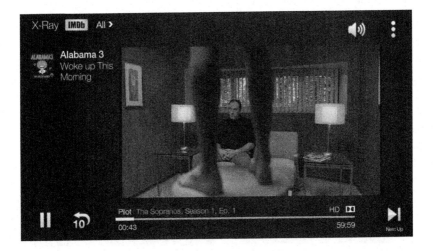

For more options, tap the menu button at upper right. From here you can turn closed captioning on and off, zoom in and out, and pause the video.

Some videos have Amazon's X-Ray, which is similar to the X-Ray feature for books (page 94). If they do, the video's main product screen will say "Includes X-Ray," and there'll be an X-Ray button at the top left of the screen. X-Ray provides information about the current scene you're watching, including the title of the music playing, the names of actors playing the characters onscreen, trivia, and so on. Tap to get more information—for example, tap an actor's name to find out more information about him.

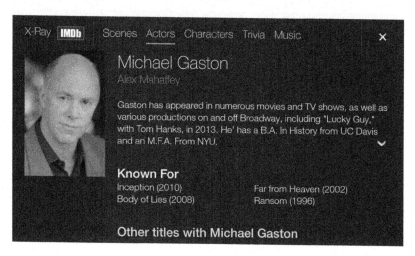

Downloading Prime Video

Some Instant Prime TV shows let you download them to your Fire phone, so you can watch them from there instead of streaming them. If a video is available for download, you'll see a Download button to the left of the Play button on the show's main screen. Tap it, and you'll come to a screen that gives you file quality options—Best, Better, and Good. The higher the quality, the larger the file, so the more space it will take up on your Fire phone and the longer it will take to download. Make sure you're connected to a WiFi network before downloading for the fastest download speed. (On average, a 60-minute video will take about four minutes to download at Best quality, two minutes for Better, and a minute-and-a-half for Good.)

If you want to check the progress of the download, swipe or swivel to pull down the Quick Actions panel. You'll see the name and thumbnail of the video and an indicator showing you what percentage of the video has been downloaded.

After the video has been downloaded, you can watch it by tapping the notification or by launching Instant Video, displaying the left pane, tapping Your Downloads, and then tapping the video you want to watch. If the video is a TV show rather than a movie, when you tap it, you won't be sent directly to the episode you downloaded—you instead get sent to the main screen of the show. Scroll down to see the episode you downloaded, and then tap it.

More Instant Video Options with the Left and Right Panels

To get the most out of Instant Video, don't forget the left and right panels—particularly the left panel. The left panel makes it easier to find videos you might be interested in viewing, because it includes a variety of categories, such as Top TV Shows, Top Movies, and For the Kids. It's also where you'll find your Watchlist and videos you've downloaded. You can also see any videos stored in the cloud and download them; tap Cloud next to Your Video Library. Then watch or download any videos you want.

NOTE The left and right panels don't work when you're watching a video. They work only when you're browsing or looking at a video's information screen.

The right panel is less helpful—it just shows recommendations for movies and TV shows, based on your viewing habits.

Using the Fire Phone with Amazon Fire TV

THERE'S ONE MORE THING your Fire phone can do with Instant Video—play the videos on a TV using Amazon Fire TV or another Instant Video-compatible television set. Amazon Fire TV is a small box you connect to your TV to watch

Prime Instant Video or play games. It also lets you use other streaming video services such as Netflix and Hulu Plus.

NOTE The Fire phone can play Instant Video on TVs in other ways as well—via game consoles such as the Xbox 360s connected to a TV, and some Blu-ray players and TVs without any additional hardware. Check the device's manual to see if it has this capability, and go here for a list of devices and more information: *http://amzn.to/ZMt3Ls*.

The Fire phone streams video to TV by using an Amazon technology called Second Screen. First, make sure that the compatible TV is turned on and connected to the Internet. Then, on the Fire phone, go to a video's description page and tap the Second Screen icon. A window appears where you can select the TV or device on which you want to watch the video. The video will start playing.

You can use the Play, Pause, and Jump Back buttons and the progress bar on your phone to move forward or back in the video. When you're done watching, tap the Second Screen icon on the Fire, and then select your Fire phone from the devices listed in order to stop playback on the TV.

Playing Music on Your Fire Phone

THE FIRE PHONE OFFERS two ways to play music—the built-in Music app or Amazon Prime Music. With the Music app, you can listen to music stored on your Fire phone or in your Cloud Drive (page 132).

Prime Music is part of the Amazon Prime subscription you get free for a year when you buy a Fire phone. You can listen to millions of free songs in Prime Music, although others you have to pay for. (Prime Music requires a separate app, as you'll learn on page 133.)

TIP Music that you've recently played shows up as individual tracks, or entire albums and playlists on the Carousel. Just tap the thumbnail of the song, album, or playlist to play it.

Using the Music App

The Music app lets you play music on your phone or in the cloud. You can get music onto your phone in several ways, including copying files from a computer (page 141), downloading them from your Amazon Cloud account (page 129), or buying them from Amazon (page 131).

If you copy music from your computer to the Fire phone, make sure to put the files in the Music folder. (It's one of the folders you see when you connect your Fire phone to your computer as described on page 144.) The app can play a variety of file types, including .mp3 and .m4a formats, and .wma, .ogg, .wav, Apple Lossless, .aiff, and .flac.

Launch the Music app from the Carousel or Apps Grid, and you'll see your music collection. To play an album or a track, tap it. If the music is stored in the cloud, and you want to download it, so you can listen to it from the phone, tap the Download icon at top right. At the bottom of the screen, you see the usual controls for playing and pausing music and moving to the next track or the previous track. If you want to play your music in random order, tap the Shuffle icon at bottom right.

If you can't get enough of the current album, playlist, or song, tap the Loop button at lower left. It starts out as two looped arrows pointing to one another. Tap it, and it changes from white to orange—that's your indication that the album or playlist will keep repeating. Tap it again, and the icon changes to a number 1 with a loop around it, signifying that the *track* you're currently playing will keep repeating. When you've had your fill, tap it again, and you get back to the original white button.

Here's a quick way to find music you want to listen to in your music collection: Tap the Search button and then type the name of an artist, track, or album. From the results, tap the music you want to play.

The Music App can play either music on your device or music you have stored in your Cloud Drive. When you play music from the cloud you stream it, although, if you want, you can download any track from the cloud to your Fire phone and then play it from there. Tap the Download button that appears at the top right when you're viewing or playing a track in the cloud. When viewing a playlist, you can tap and hold a song on the list to download it. You also get options to view lyrics, add to playlist, explore artist, go to album, or delete from Cloud.

To switch back and forth between music stored in the cloud and music on your phone, open the left panel, and then tap either Cloud or Device.

Creating and editing playlists

A music player without playlists is a sad thing—so, of course, the Fire phone lets you easily create playlists. With a playlist, you become the disc jockey of your own radio station, because you can play a collection of songs in exactly the order you want. First, display the left panel, tap Playlists, and choose Create New Playlist. You must name the playlist and save it before you can add music to it. After you save, you'll see a list of all your music in the cloud as well as on the Fire phone. Tap the + sign next to the tracks you want to add, and then tap the checkmark at the top of the screen.

To play any playlist, display the left panel and tap it. Use the same control for playing an album or track as you do for the playlist. To edit the playlist, tap the pencil icon at the top of the screen. To delete tracks, tap the – button next to any. To add tracks, tap the + at the top of the screen and add them the usual way.

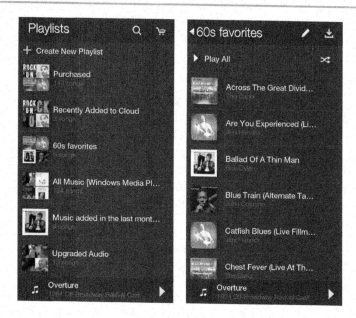

More options with the left and right panels

When you're in the Music App, the left and right panels are your friends—especially the left panel. The left panel gives you multiple ways to view your music collection: by artists, albums, songs, and genres. As detailed previously, it's also where you get to your playlists. From here you also can buy music (page 131).

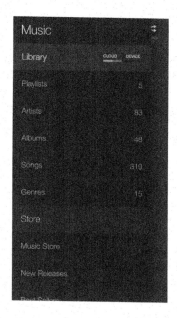

The right panel does double-duty. When you're not listening to music, it displays new music releases and best sellers. When you're listening to music, it displays the lyrics courtesy of the X-Ray for Music feature, if it happens to have those lyrics in its database. It turns the right panel into a little karaoke machine. The words currently being sung are highlighted as the song progresses. You can scroll up and down through the full lyrics, tap any, and the song will jump to the tapped lyric.

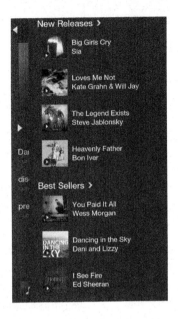

Buying music from the Music app

Amazon sells everything under the sun and likes making it easy for people to buy it all. Not surprisingly, you can buy music from Amazon right from the Music app. When you're in the app, tap the Store button at the upper right of the screen. When you arrive at the Music store, you can search and browse. Tap the song or album you're interested in, see details about it, and then buy it if you're interested. It gets downloaded to your Fire phone and is also available via the cloud.

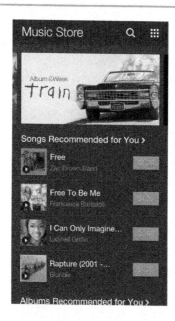

Fire Radio

Another way to play music by using your Fire phone is to use an app that turns the phone into an FM radio or streams music to the phone wirelessly. To use your Fire phone as an FM radio, head to the Amazon Appstore and download Tune-In Radio, which streams FM radio stations from all over the world.

Another great bet is Spotify, which lets you stream individual tracks to your Fire phone. At last count, it had more than 20 million songs on it.

Playing Music in the Cloud

YOU CAN PLAY MUSIC not just on the phone, but music you've stored in your Cloud Drive as well. You know how music ends up on the Fire phone—you copy it there, download it, or buy it. But how does it end up in the cloud? Either of two ways. If you have other Amazon devices with music on them, that music is copied to the cloud automatically, and you can then play it from the cloud using the Fire phone, or download it to the phone.

There's another way as well, via the Amazon Music web page. It lets you upload music from your computer and then play it wherever you are on the Web. When you upload music to it, you're uploading the music to your Amazon cloud, and the music is then available via the cloud on the Fire phone, as well as via the Web.

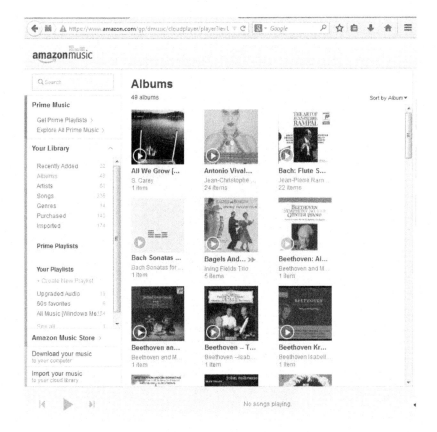

To get started with Amazon Music, go to this link: *http://amzn.to/1w6t4HZ*.

Using Prime Music

WITH YOUR SUBSCRIPTION TO Amazon Prime, you get to listen to free music, and plenty of it—more than a million songs. To use it on the Fire phone, you have to download the Amazon Prime Music app from the Amazon Appstore (page 259). You can't listen to Prime Music using the built-in Music app. Go figure.

The Prime Music app, as you'll see in this section, is a bit confusing to use. And it's not just the app that's confusing, but understanding the difference between Prime Music and your Prime Music library. Your own library includes music that you've uploaded to the cloud as detailed in the previous section, music that you've copied to your phone, music you've bought, and music from Prime Music that you've added to your library. The Prime Music app lets you play music from Prime Music, music that you've added to your library from Prime Music, as well as other music you've uploaded to your library.

NOTE Each time you play a track in Prime Music, it gets automatically added to your library. For details about another way of adding music from Prime Music to your library, see the next section.

Finding Music in the Prime Music App

Launch the Amazon Prime Music app, and you'll see songs listed by song, album, or playlist, which have already been created for you. Near the top of the screen you'll find navigation buttons. You can browse by Songs, Albums, or Playlists. As you browse songs, you'll see that some of them have an Add button next to them, and others have a button that says "In Library." Tap Add to add it to your library. Be sure to tap the Add button rather than the Menu button (three stacked dots) to its right. If you tap Menu, you'll instead have the option of *buying* the track.

NOTE You can only listen to music you've added to your library from Prime Music for as long as you subscribe to Amazon Prime. Once the subscription lapses, you won't have access to that music anymore. However, if you buy the track, you'll *always* have access to it.

Browsing Prime Music by genre

When you're browsing tracks, songs, and playlists, you may not want to see the full Prime Music selection. You may only be interested in Country, Alternative Rock, Classical, or another genre. Tap the small right triangle at the top middle of the screen, and from the screen that appears, select the genre you're interested in.

Searching Prime Music

You can also search for tracks, artists, and albums using the Search box at the top of the screen. When you look for music, you may notice something odd—not everything is free. That's because the Prime Music app also finds for-sale tracks and albums in addition to those that come free with your Prime subscription. If you want to see only free tracks and albums, tap the disclosure triangle at the top of the screen, and then move the slider to tell the app to display only Prime music.

Playing Music in the Prime Music App

The Prime Music app isn't organized like the built-in Music app, and it takes some getting used to. To play a track, tap it in the search results or any Prime Music screen. What happens next can get a bit confusing.

At the top of the screen, you'll see the *album* that the track is from, not the track itself. If you want to play the track itself, you have to look about halfway down the screen, where you'll see the track listing, along with all the other tracks from the album. Tap any track you want to listen to, and it starts playing.

On the rest of the screen, you have an option to buy the entire album, even though it's available for free from Prime Music. The difference is that buying the album downloads it to your Fire phone, rather than streaming it. You'll also see options for adding the album to your library (page 133). When you do that, recommendations for other albums appear at the bottom of the screen.

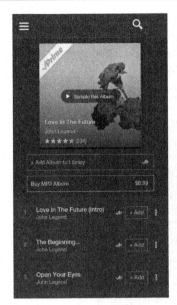

The player has buttons for playing, pausing, and moving forward or back at the bottom of the screen. For more controls, such as loop shuffle and changing the volume, tap the Menu icon (three stacked dots) at top right.

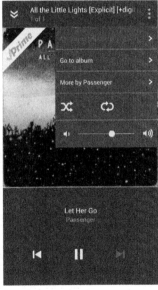

To get more of these controls onscreen, tap the two arrows at the top left. The app reconfigures itself, making the album picture smaller and adding buttons for looping and downloading. To see the rest of the features, including changing the volume, browsing to a different album, and seeing more tracks by the same musician, tap the Menu button at top right.

Using the Left Panel

Swipe or tilt to open the left panel for more ways to browse Prime Music, including by recently played, recently added, and so on. You can also access the Music store.

The right panel doesn't work in the Prime Music app. That means that there's also no X-Ray feature showing you a song's lyrics, as there is in the built-in Music app (page 131).

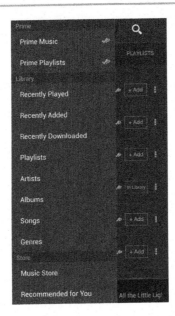

All Docs   Q

Add Docs to Your Cloud Drive

✉ Email Docs to your Device

☁ Sync with Cloud Drive

🌐 Clip articles from the web

▢ Browse Docs on your Device

You'll learn to:

- Transfer files to your PC or Mac

- Browse through your files with an Android app

- Check available space on your Fire

- Sync files with the cloud

CHAPTER 6

Transferring Media and Files, and Syncing with the Cloud

YOUR FIRE PHONE IS not an island—it's built to work with computers, so you can get your music collection, photos, and videos from your main machine onto your phone. In fact, you can transfer just about any type of file between your computer and your Fire. The Fire can also grab files from the cloud—books, games, music, videos, photos, and documents.

Connecting Your Fire to Your Computer

TO TRANSFER FILES BETWEEN your Fire and your computer, first connect your Fire phone to your Windows or Mac computer by using the phone's micro-USB cable. Connect the micro-USB plug into your Fire phone and the normal-sized USB plug into your computer's USB port.

NOTE When you connect your Fire to your Windows PC for the first time, the PC may not recognize it. It may need a special *driver*—a small piece of helper software—to see the Fire and communicate with it. Your PC will look for and install the drivers automatically (as long as it's connected to the Internet).

When you connect it you'll hear a small audio notification. That means the connection has been made. Swipe or swivel to pull down the Quick Actions panel, and you'll see a notice that reads "Connected as a media device." Tap the notification, and the USB Connection screen tells you you've connected your phone to your computer for file transfer—you'll see the checkbox turned on. (For the techies, the Fire phone connects using Media Transfer Protocol.)

Underneath that setting, there's another one—Photo Transfer. If you're transferring photos from a Windows computer, you can ignore this setting. You'll grab photos from your computer just like any other files, using Windows Explorer, as described in the next section. With a Mac, you need a little help from a program called Android File Transfer to move files to the Fire phone. Turning on Photo Transfer lets you use iPhoto to move your photos to the Fire phone instead; for the details, flip to page 145.

Transferring Files from a Windows PC

WHEN YOU CONNECT YOUR Fire phone to a PC (and the drivers are installed, if necessary), the AutoPlay screen may appear, just as when you connect any USB device. However, don't be concerned if you don't get this screen, because it doesn't always show up. It all depends on your Windows version and computer setup (and maybe even the phase of the moon).

After you've connected the Fire phone to your PC, launch File Explorer (in Windows 8) or Windows Explorer (in any other version of Windows). Your Fire phone shows up as a removable disk—surprisingly, called Kindle, not Fire. (It's an Amazon thing.)

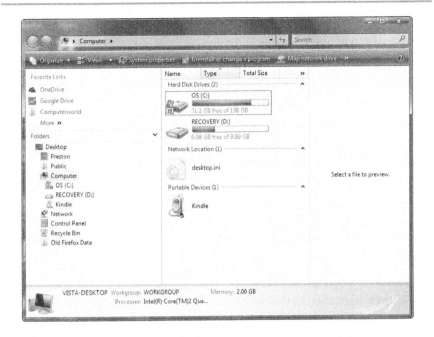

You can now use your phone as if it were any USB flash device—copying files to and from it, creating folders, and so on. You just need to learn your way around. You'll find a number of folders on the Fire, some of whose names are clear, and some of whose names make no sense—like DCIM.

There are several important folders that contain information you might want to transfer from your Fire to your PC, or vice versa. You'll see a lot more folders, but these are the important ones:

- **Amazonmp3.** Here's where the Fire stores music that you've downloaded from the Amazon MP3 store (page 131).

- **Documents.** If you've got Word files, PDFs, and similar files, here's where they'll be.

- **Download.** This is a confusing folder, and one you may never use. It's not the same as the Downloads folder on your Fire phone that's used to store files you've downloaded from the Internet, such as pictures or web pages. Instead, it's a folder that you can use to transfer files to and from your PC or Mac.

- **DCIM**. Here's where the Fire phone stores all the photos and videos you've taken. Drag photos from this folder to your PC to copy them, or drag photos from your PC to here to put them into the Gallery (page 105).

- **Movies**. As the name says, here's where you'll find movies and videos you've transferred to your Fire phone.

- **Music**. The Fire phone stores music here, although it might also store some in other places as well. If you download music files by using the Amazon music app, for example, there will be an Amazonmp3 folder.

NOTE When you use a PC or Mac to view the files on your Fire phone, you won't see every folder. Buried deep in the phone's file system are important files like books you've bought or gotten for free from Amazon. The Fire phone hides these folders so you can't mess with them.

- **Pictures**. Here's where to drag photos when you want to copy them from your PC or Mac. Photos you take with the Fire phone's camera are instead stored in the DCIM folder.

- **Podcasts**. Note that podcast apps, such as Podcast Addict, don't download to this folder. Instead, this folder lets you transfer podcasts from your PC or Mac to the Fire phone and remember where you put them.

- **Ringtones**. Downloaded ringtones? Here's where to find them. (See page 73 for information about downloading ringtones.)

If you're transferring music from your PC to your Fire phone, either copy it into the Music folder or create a new folder on the phone using File Explorer or Windows Explorer, as you normally would. Then transfer the music to that folder, using subfolders if you want. Your Fire phone's music player (page 127) will automatically recognize the music you've transferred. The same holds true for books and movies—simply transfer them to the appropriate folder on your Fire.

Transferring Files from a Mac

TO TRANSFER FILES BETWEEN your Mac and Fire phone, you must first download the free Android File Transfer tool. Go to *www.android.com/filetransfer/* and follow the instructions for downloading and installing it.

Once you do that, connect the Fire phone to your Mac with the USB cable. Then run the Android File Transfer app. After you run it for the first time, from then on it should automatically start whenever you connect your Fire phone to your Mac. You can then transfer files back and forth between your Mac and your Fire phone using Android File Transfer. It works just like Finder or any other file management app. You'll see the same file structure as outlined on page 143.

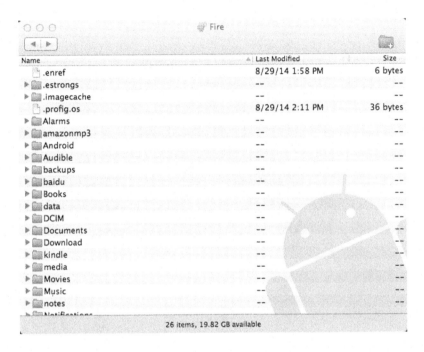

If you want to transfer photos specifically between your Mac and Fire phone, you can use iPhoto to do it. First, connect your Fire to your Mac via USB. Then pull down the Quick Actions bar and tap the notice that reads "Connected as a media device." On the USB Connection screen, turn on the Photo Transfer checkbox. iPhoto will launch and show all the photos on your Fire phone. When iPhoto opens, click Import 100 Photos (or whatever number is on your Fire phone), or select only the photos you want to import into iPhoto and then click Import Selected. If you have an older version of iPhoto, when it opens, you can choose either Import All or Import Selected.

Depending on the programs you have on your Mac and their settings, iPhoto may not open automatically when you plug in your Fire phone. A program called Image Capture may open instead. To use iPhoto, launch iPhoto and go to iPhoto→Preferences →General. Next to "Connecting camera opens," choose iPhoto. (Unless you use iPhoto as your primary photo-organizing program, though, you may find it easier to stick with Android File Transfer and drag photos from your Fire phone to your Mac and vice versa just like any other files.)

Using Android Apps to Manage Your Files

TO BROWSE THROUGH THE files of your Fire phone, you don't need to rely on your PC or Mac—you can use any of several Android apps. You won't use these apps for transferring files between your computer and your Fire. Instead, you'll use them to browse through files, and delete and move any, if you so desire.

Moving files around can be dangerous. You may delete important files or some of your Kindle books. Only move or delete files if you know what you're doing technically.

One benefit of these types of apps is that they typically show you all the folders on your Fire, something that you won't see when you connect the Fire to your PC or Mac. So you'll be able to see the *sdcard* folder and all the subfolders underneath it, which you can't do on the PC or Mac.

The Amazon Appstore has several good free apps for browsing through your files, including ES File Explorer and Ghost Commander. To find others, search for "file manager" in the Appstore.

 ## Checking Space on Your Fire

IF YOU TRANSFER LOTS of music and files from your PC or Mac to your Fire, you may eventually run out of storage space. It's a good idea to regularly check how much space you've got left. Pull down the Quick Actions panel and select Settings→Battery & Storage→View Available Storage.

You'll come to a screen that shows you how much storage your files take up and how much storage you've still got available on the Fire.

If you're running out of space, there's a way to get some quick extra storage. Tap Clear Storage Space, and the Fire phone will delete any items you haven't used recently. But first, it shows you what it plans to delete so you can uncheck the boxes next to anything you don't want deleted. After making your decision about what to delete, tap the Trash icon at the bottom of the screen to delete the selected files.

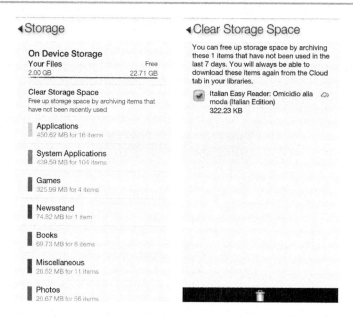

You can also clear storage space by deleting apps (page 267), and deleting files and media content by using a file manager (page 146) or connecting your Fire to your PC or Mac and using the computer's file manager.

WARNING If you're deleting music, use the Music app rather than the Files app. If you delete the files using the Files app, the Music app may still show the music as being present, even after the files are gone.

Downloading from the Cloud

THE FIRE PHONE IS designed to work with your Amazon account and with the cloud. To be specific, it uses the Amazon Cloud, which stores the things you buy from Amazon. Anything you buy is automatically saved to the Amazon Cloud. And that means that it's easy to get that content from the cloud to your Fire phone. To download the content, from the Carousel or Apps Grid, swipe or tilt to open the left panel, and then tap any of the following categories:

- **Music.** Tap here, and then tap Cloud to see any music you've stored in the cloud, for example, in the Amazon Cloud Player (page 132). You'll be able to browse your music by playlists, artists, albums, songs, and genres. Tap any song to stream it and listen to it. Hold your finger on any song for a list of actions, including downloading it to your Fire phone.

- **Videos.** Tap here to see any videos you have stored in the cloud, for example that you've bought from Amazon but haven't yet played on your Fire phone. You can download or stream them to the phone from here.

- **Books.** Shows you books on your Fire as well as books in the cloud.

- **Docs.** From the screen that appears when you tap this option, tap "Sync with Cloud Drive" if you want to upload documents on your Fire to the cloud, and download any from the cloud onto your Fire.

NOTE For information about how to download apps and games from the cloud to your device, turn to page 263.

The Fire Phone Online

◀ Wi-Fi

Wi-Fi networks.

Connect to a Network

📶 CoxWiFi

📶 xfinitywifi

📶 🔒 The Bureau of Reclaimed Spa

📶 Google Starbucks

📶 CableWiFi

📶 🔒 lesley_secured

📶 🔒 5JNJR

You'll learn to:
- Connect to a WiFi network
- Turn your Fire phone into a WiFi hotspot
- Use Airplane mode

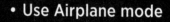

CHAPTER 7

Getting Online: WiFi, 4G, and Mobile Hotspots

THE FIRE PHONE WAS built as much for going online as for making phone calls. It gives you access to the Web, maps, videos, and more. Just about anything you can do on the Internet on a computer, you can do on your Fire phone.

But first you have to get online. This chapter shows you the ways you can do so, plus a few more tricks like turning your Fire into a WiFi hotspot to provide an Internet connection for nearby computers and other devices.

How the Fire Phone Gets Online

THE MOMENT YOU FIRE up your Fire, you're ready to go online. You're automatically connected to a high-speed 3G or 4G connection. These are called *data connections*, and they're designed for getting on the Internet fast.

NOTE When the 3G or 4G network isn't available, the Fire phone still lets you make calls using an older, slower network. But that network isn't suitable for web browsing or other Internet activity.

To connect via 3G, 4G, or the voice network, there's nothing you need to do. Your Fire connects automatically. WiFi, the fastest kind of connection, takes a little more work, as you'll see in the next section.

Connecting via WiFi

WHEN YOU CONNECT TO the Internet via a WiFi hotspot, you usually get a very fast connection. If you get a fast hotspot, your speed can be as fast as your cable modem connection at home.

Hotspots are everywhere. They're in coffee shops and hotels, which frequently offer free WiFi, although some make you pay for it. You'll find WiFi coverage in airplanes, libraries, and even entire cities. Some cable companies also have WiFi hotspots, so check with yours. If you connect your computers to the Internet at home using a wireless router, you have your own WiFi hotspot, and you can connect your Fire to the Internet via your home network.

NOTE When you're connected to a WiFi hotspot, the Fire phone uses it for more than just Internet access. It also uses WiFi for finding your current location in apps like Google Maps—with or without help from GPS, depending on your settings. With WiFi, your phone finds nearby WiFi networks and uses fancy algorithms to determine your location. It's not as precise as GPS, but it's still pretty good.

Your connection speed will vary from hotspot to hotspot and can even change at an individual hotspot depending on how many other people are using it. If a lot of people are using it at once, and the hotspot isn't set up to handle that many connections, your speed may suffer. Also, keep in mind that WiFi is not the way to connect to the Internet if you're in a moving car. Hotspots generally have a range of about 300 feet, so you and your phone may zip right past them.

NOTE You've probably seen some new car models advertised as having WiFi hotspots built into them. What these cars actually have is built-in 4G/LTE access to a cellular network, just like your Fire phone has. They can then share that access with devices like tablets and phones via WiFi. All you have to do is turn on your Fire phone (or other device) and connect to the car's WiFi just like any other WiFi network described in this chapter.

Unlike the 3G/4G connection, which happens automatically, WiFi doesn't work unless you turn it on. The fastest way to turn it on is to use the WiFi widget. To get to it, swipe or swivel to pull down the Quick Actions panel and tap the WiFi icon. It turns orange to show WiFi is on. If you've previously connected to a network, and your settings are all aligned, you'll automatically connect to one of your preferred WiFi networks, like the one you use at home. If you've never connected to a network, or you're not near one you've used before, you need to find a network.

WiFi icon

To connect to a WiFi network for the first time, pull down the Quick Actions panel and tap Settings. Then tap Wi-Fi & Networks→"Connect to Wi-Fi." You'll see a list of all the WiFi networks near you. Tap Scan to tell your Fire to look again for a network, in case it missed one you believe is nearby. A list of all nearby networks appears. For each network, you'll see its name, as well as the relative strength of the connection: the more waves, the stronger the connection. In addition, some networks have lock icons on them. A lock means the network is encrypted and password-protected, so you'll need the password to connect to it.

On the screen that lists all nearby WiFi connections, you'll also see a list of networks you've connected to in the past, even those out of range. Those networks don't show a WiFi icon, and you can't connect to them.

Tap any network on the list. If it's open and doesn't require a password, you'll connect right to it. If it's protected, you'll come to a screen that requires a password; type it and tap OK. After a few seconds, you'll see the word "Connected" underneath the network's name. To see details about your WiFi connection, tap the name. You'll come to a screen that tells you your signal strength, speed (what the Fire calls Link Speed), type of security, and your IP address. From now on, whenever you're within range of the network, the Fire phone will connect to it automatically if WiFi is turned on.

NOTE In some instances, even though a network shows that it doesn't require a password, you'll need a password in order to use it. In those cases, you can connect to the network without typing a password, but you'll have to open your browser and register or type a password. Often, these kinds of networks are for-pay. For details, see page 158.

◄Gralla

Signal Strength
Excellent

Status
Connected

Link Speed
54Mbps

Security
WPA/WPA2 PSK

IP Address
192.168.1.117

Forget This Network

What if the screen showing you all the available nearby networks doesn't appear? There may be no networks within range...or it may just be some odd temporary glitch. To check whether there are any available networks nearby, go to the Wi-Fi Settings page to see the screen that shows you all the nearby WiFi networks so you can make a connection.

TIP If your cellphone plan includes a limit on the amount of data you can use every month, and charges you more if you go over the limit, the Fire's WiFi capabilities can be your best friend. When you're sending or receiving data via WiFi, it's not counted as part of your data plan. So using WiFi whenever possible can help make sure you don't bump up against your limits. And if you know you're going to be downloading large files, such as songs or movies, try to do it via WiFi. In addition to not racking up data use, it'll be faster than 3G or 4G.

Disconnecting and Reconnecting

To disconnect from a WiFi network, turn off WiFi. If you want to keep WiFi on, but want to disconnect from the network, go to Settings→Wi-Fi & Networks→"Connect to Wi-Fi." That lists all the nearby WiFi networks. Tap the network to which you're connected, and then tap Forget This Network. Boom— you're disconnected. (You can also forget a network using the same method even if you're not connected to it.)

There's a downside to disconnecting this way, though. Normally whenever you connect to a WiFi network, the Fire remembers that connection, so the next time it's in range, the phone automatically connects you—you don't even have to re-enter the network's password. If you tap Forget This Network, though, the Fire won't log you in automatically the next time you're in range.

Connecting to For-Pay WiFi Networks

Some WiFi hotspots require you to pay a fee for their use. First make the connection in the normal way. Then launch the Silk browser by tapping its icon on the Carousel or the Apps Grid. A screen appears, delivered by the network, asking you to first register and pay.

Some free WiFi networks require you to agree to terms of service before you can use them. In that case, when you launch the browser, those terms of service will appear. So if you've connected to a free WiFi hotspot but can't get an Internet-based app like Pandora to work, it might be because you haven't yet launched your browser and agreed to the terms of service.

Connecting to an "Invisible" Network

For security reasons, some people or businesses tell their network not to broadcast its name—its *Service Set Identifier* (SSID). That way, it won't appear as an available network. (Dedicated hackers, though, can easily detect it.)

If you need to connect to a network that isn't broadcasting its SSID, you can do so, as long as you've been provided with its name, the type of security it uses, and its password. Go to Settings→Wi-Fi & Networks→"Connect to Wi-Fi"; on the next screen, scroll to the bottom and tap Add Network. Type the network's SSID, choose the security type, type the password, and then tap OK to connect to the network.

Turning Your Fire Phone into a WiFi Hotspot

THE FIRE PHONE CAN do more than just connect you to a hotspot. It can create its own hotspot, so other computers, cellphones, and devices can connect to the Internet through it. That means, for example, that if you've got a computer that you want to connect to the Internet but there's no WiFi hotspot or Internet service nearby, you can connect using your phone. In fact, you can provide Internet access not just for one device but for multiple ones. (The exact number you can connect may vary according to your carrier, so check with it for details.)

NOTE If you don't have a 3G or 4G connection, you won't be able to set up a mobile hotspot. So try doing this only when you see the 3G or 4G signal in the Status bar.

To perform this magic, the Fire connects to a 3G or 4G network as it normally does, using its 3G or 4G radio. Then it uses its WiFi radio to set up a WiFi hotspot, and lets multiple computers, phones, tablets, and similar devices connect to it, just as they would to any other hotspot. Don't expect blazing speed if several devices are using this single 3G or 4G connection simultaneously. Still, it's a high-speed connection.

Be aware that this hotspot may come at a price. Depending on your provider and its current plans, you may have to pay an extra $20 per month or so in addition to your normal data fee to be able to use this feature. And there may be maximum data limits imposed as well. That will apply only to data sent and received via the 3G or 4G hotspot, not toward your normal data plan.

When you're ready to set up a hotspot, first make sure WiFi is turned on and you've got a 3G or 4G connection. Then go to Settings→Wi-Fi & Networks→"Set up a Wi-Fi hotspot."

On the next screen, tap Configure Hotspot. You've got some techie details to fill in, but don't despair—you won't have to fill them in every time you want to set up a hotspot. Do it just once and you'll be all set.

Here's what you'll fill out:

- **Network SSID**. Type the name that you want your hotspot to have. The box will be filled in for you already, with something really exciting like "Fire_7672." You can make the name anything you want—just make sure it's something you'll remember.

- **Hide your network name (SSID) from being visibly broadcasted**. With this box unchecked, anyone can see the network's name, because it's being broadcast, which is the way networks normally work. But if you're paranoid, turn this setting on. No one will be able to see that your network exists. You, however, can still connect to it—for details, see page 158.

- **Power Management**. This setting determines how long the hotspot should stay active even if no one is connected to it. If you're worried about battery life, pay attention to this one. Hotspots can suck up a lot of battery juice, so you don't want it active for too long if no one is using it. Out of the box, it's set to 10 minutes. But you can also set it to have it stay on for only 5 minutes, or to always stay on.

- **Security**. This drop-down menu lets you choose the type of security you want your network to have. WPA2 PSK will be chosen for you. Your other choice is Open. Resist any impulse to choose Open; if you do that, anyone can connect to your hotspot. Not only will they suck up your bandwidth, but they could also possibly steal your files.

- **Password**. Select the password that you (or anyone else) will have to type in in order to connect. Please, whatever you do, don't use the word "password" as a password. For better security, use a mix of numbers and letters, including capital letters. You'll have to make the password at least eight characters long.

Tap Save, and you're ready to go. Keep in mind, though, at this point you actually haven't turned on your hotspot—you've merely configured its settings. To turn it on, after you tap Save, turn the slider at the top of the next screen to On. A message appears at the top of the screen telling you the hotspot is on, and for how long. You'll also see the number of devices connected to it. Also, when your hotspot is active, there'll be a notice on the Lock screen telling you so.

When you're ready to shut down the hotspot, get back to the screen where you set it up, and then flip the On switch to Off.

Airplane Mode

TIME WAS, AIRLINES REQUIRED all passengers to turn off their cellphones' network connections—voice, data, and WiFi—during flights. That's where Airplane mode comes in. It turns off all your Fire phone's radios but lets you use your phone for games, apps, and so on. To turn on Airplane mode, go to Settings→Wi-Fi & Networks→"Enable Airplane mode." On the next screen, in the Airplane Mode section, move the slider from Off to On. Go back to this screen when you're ready to turn Airplane mode off.

NOTE Don't let the term "Airplane mode" limit you. You can also use this mode when you're not in flight in order to save power.

While all airlines make you turn off your radios during takeoff and landing, the rules may vary when you're in the air. Some airlines let you use WiFi, for example, and even offer onboard WiFi hotspots. A few international airlines have even started letting passengers make voice and data calls, and more may follow suit. Airplane mode is keeping up with the times; you can turn on Airplane mode and then turn on only WiFi or voice access.

NOTE If your airline allows WiFi but not voice or SMS text messaging (page 76), you can still exchange text messages with your friends. Just use a WiFi-based texting app like WhatsApp, Kik, or Facebook Messenger.

More Network Settings

THE MOBILE NETWORK SCREEN, where you turn Airplane mode on and off, gives you several other useful settings:

- **Cellular Data.** This option is normally turned on, so you can access your provider's network for cellular data—that is, the one for web browsing, using apps, and the Internet in general. Why would you ever turn it off? If you travel overseas, you'll typically pay outrageous amounts of money for data access, so you may want to turn it off before traveling.

- **Data Roaming.** This setting determines whether you can access data from a different network than your normal provider's when you're outside its cellular data range. When it's set to Off, you can't access data from another network. Again, this is one you might want to make sure is turned off before traveling overseas.

- **Data Usage.** Shows how much cellular data you've used for the month and which apps are using that data. It's important to keep an eye on this number if you have a data plan that limits the data you can use and that charges you extra if you exceed it. Tap Manage Data Limit for a variety of settings, such as setting an alert to fire when you're approaching your data limit and turning off cellular data after a certain limit.

theonion.com

the ONION®
America's Finest News Source

Date's Flaws Coming At Woman Faster Than She Can Rationalize Them

Going-Out-Of-Business Sign Thanks Neighborhood For 3 Months Of No Support Whatsoever

‹ › 9

You'll learn to:

- Browse the Web
- Use multiple browser tabs
- Create and manage bookmarks
- Save and view online pictures and graphics

CHAPTER 8

The Web

WITH THE FIRE PHONE, you've got the whole Web in your hand. Whether reading reviews and buying movie tickets, reading online articles while you commute, or checking out YouTube, you'll find the Fire to be a great web browsing companion. The 4.7-inch screen gives you a superb web experience wherever you go. With more and more web designers making their sites look good and work well on mobile devices like the Fire, you may find yourself using the browser more than any other app.

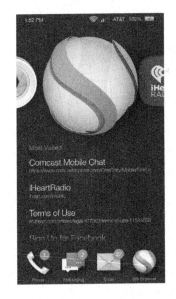

Silk—The Fire Phone's Browser

THE SILK BROWSER OFFERS plenty of the same goodies as a desktop browser, including bookmarks, auto-complete for web addresses, saving and sharing pages, and more. However, the browser itself is fairly bare bones, which may take some getting used to. But once you learn your way around, you'll be browsing at warp speed. Here are the main controls you need to know about:

Address bar   Menu

Back button   Forward button   Open tabs

- **Address bar**. Here's where you enter the URL—the web address—for a page you want to visit.

> **NOTE**   Silk's address bar does more than let you enter a URL. It's an *omnibar*, since it also lets you search the Web. Just type your search terms right into the bar. See page 174 for details.

- **Open tabs**. Tells you how many websites are currently open in a tab. Tap it to see thumbnails of them all. Then tap any thumbnail to visit it. Tap the X to close any site. Tap the + at the top of the screen to open a site in a new tab.

- **Left panel**. On this panel, you can access a variety of features— your bookmarks, saved pages, downloads, and more. See page 183 for more details.

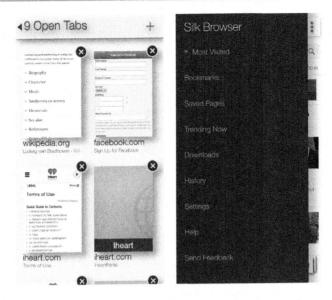

- **Right panel**. The right panel lists a table of contents of the current website (page 187). Tap any link to jump straight to it.

- **Forward and back buttons**. Tap to move either forward or back in your browsing session.

- **Menu**. Tap the three vertical dots to bring up a menu that lets you perform a variety of functions, such as sharing the pages, adding a bookmark, and more (page 182).

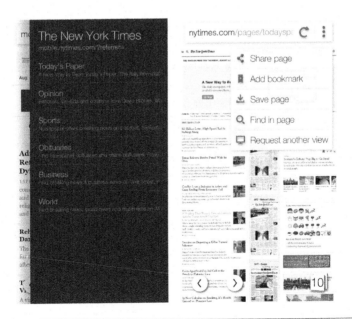

As of this writing, Silk is the only browser that works on the Fire phone; you won't find any old favorites like Chrome or Firefox in the Appstore. However, it's possible that Amazon may make other browsers available someday, so keep checking.

You may notice something confusing as you use Silk. Sometimes you'll see the "Open tabs" icon and the forward and back buttons, and sometimes you won't. It may seem that they appear and disappear at random, but there's nothing random about it. It is, however, an odd design choice. If you scroll toward the top of a web page—by moving your finger down the Fire phone's screen or by tilting the phone toward you—these tools appear. Scroll toward the bottom of the page, and they disappear. The address bar also disappears when you scroll to the very bottom of a web page.

The Fire's Dynamic Perspective navigation features (page 29) work in Silk. For example, you can scroll up and down web pages by tilting the phone away from you or toward you, make the right panel and left panel appear by using the Tilt gesture, and so on. You can, of course, also use traditional gestures like swiping.

Basic Navigation and Managing Tabs

JUST AS ON A desktop browser, your phone's Silk browser lets you open multiple tabs and visit multiple sites—one on each tab. When you're in the browser, tap the icon on the lower right with a number on it—that number displays the number of tabs open. You'll then see all your open tabs. Tap any to go to it; tap the X to close it. And tap the + at the top of the screen to open a new window.

When you open a new window, you'll come to a page that shows your most visited sites. Tap any to open it. You can also type a URL in the address bar to visit a website, or type in a search term to search the Web. And when you tap Open Tabs, you get sent to the tabs that you've most recently opened.

Navigating a Web Page

WHEN YOU BROWSE THE Web with Silk, some sites will appear with the same fonts, links, pictures, and so on, just as if you were visiting it on a computer with a much larger screen. Others, though, will be stripped-down, simpler versions, since they're designed for smartphones' smaller screens. (More on that in the next section.)

If you land on a page that was designed for a full-size screen, you'll find that the type is minuscule, the photos small, and the links hard to detect. But letting you see the entire screen at once makes a good deal of sense, because at a glance, you can see what section of the page you want to focus in on.

That's where the fun begins. You can use the Fire phone's zooming and scrolling capabilities to head quickly to the part of the page you want to view, and then zoom in.

You've got two ways to do so:

- **Use the two-finger spread.** Put two fingers on the screen on the area where you want to zoom in, and move your fingers apart. The web page stretches and zooms in. The more you spread, the greater the zoom. Pinch your fingers together to zoom back out. You may need to do the two-finger spread and pinch several times until you get the exact magnification you want.

- **Double-tap**. Double-tap with a finger on the section of the page where you want to zoom. Double-tap again to zoom out. You can't control the zoom level as finely with the double-tap as you can using the two-finger spread.

Once you've zoomed in, you can scroll around the web page by dragging or flicking your finger—the same kind of navigation you use for other Fire apps (page 21). And you can scroll up and down web pages by tilting the phone away from you or toward you (page 28).

Web Pages Designed for Mobile Phones

AS YOU BROWSE THE Web, you may come across sites that differ significantly when viewed on the Fire (or other smartphones) compared with the exact same sites viewed on a computer. That's because web designers have created pages specifically designed to take into account that mobile phones have smaller screens.

CNN, for example, has sites designed especially for mobile viewing. Head to the same site at the exact same time of day with a Fire phone and a computer, and you'll see very different pages, even though the content of the pages is much the same. As explained earlier, when you view a web page designed for a large screen on a small screen, the text can be too small to read and links can be too tiny to tap.

By contrast, pages formatted to be read on the phone present articles and other information in scrollable lists instead of complex layouts, to make it easy to find what you want quickly. They may not let you zoom in and out. You navigate primarily by scrolling and clicking links.

There are pros and cons to each type of design: pages designed for mobile may lack features such as photos and videos, while pages designed for larger-screened devices may be too difficult to navigate and read on the phone.

The good news: The Fire lets you easily switch between the two types of design—as long as the website offers the two versions. When you come across a web page designed for a computer rather than a phone, you can tell the Silk browser to display a phone-friendly version of the page. Tap the menu button at upper right ⋮ and select "Request another view." A screen appears with three choices: Automatic, Desktop, and Mobile. Automatic is usually the best choice, because then Silk does your work for you, deciding which view of the web page would be best. But in some cases, if you come across a page designed for a computer rather than a phone, you want to tell Silk to instead show the page designed for the phone. In that case, tap Mobile, and then tap OK. Then retype the page's URL in the address bar, and you should then see the mobile page.

TIP After you've chosen a mobile web page, though, go back to the menu and change back to Automatic, so Silk can use its smarts to make the right choice when you browse.

The Address Bar

THE ADDRESS BAR IS the box at the top of the browser where you type the URL of the website you want to visit. When you type an address and head to a page, a yellow bar appears in the address bar that shows you the status—approximately how much of the page has loaded and how much is left to go. If a page loads in a flash, you'll barely even see the yellow.

Typing an Address and Searching

To type a URL into the address bar, first tap the bar. The current URL is highlighted in blue. Then use the keyboard to type an address. As you type, the Fire displays sites you've visited that match the letters you type. So when you type the letter *C*, for example, it may display Computerworld.com (*http://Computerworld.com*), CNN.com (*http://CNN.com*), and so on. It will also display search terms you might want to use, because the address bar does double-duty as a search box. So you may see a very long list of URLs and suggestions.

> **NOTE** As you type, you'll also see suggestions for sites you might not yet have visited. Bing is trying to be helpful and displays popular sites that match the letters you're typing.

You'll notice, though, that it might also display URLs that don't start with the letter C. If you've previously visited a site about the international opera star Cecilia Bartoli, you may see that site come up when you type *C*. That's because the browser looks through your browsing history and Bookmarks list (see page 176), and looks for *all* matches to that letter, not just in URLs but also text in the page's title. When it displays its list as you type, it includes both the page's title and the URL.

As you continue to type, the list narrows down, and matches only those sites that match the letters you're typing. So if you type *com*, cnn.com (*http://cnn. com*) no longer appears in your list, but computerworld.com (*http://computer-world.com*) does. When you see the site you want to visit, just tap its listing. You head straight there. If there's no match on the list, you'll have to type the entire URL.

NOTE   Don't bother to type the *http://* part of a web address. The browser knows to put that in for you. You do, however, need to type in the .com or other ending, such as .edu. After you type in the address, tap the arrow button, and you head to the page.

You can also use the address bar to search when you don't know the name of a website. Just type your search term, but don't add a .com ending. Your browser will use Bing to search the Web for sites about that term.

TIP If you sometimes find yourself with a slow Internet connection and wish there was a way to browse the Web faster, here's a bookmark you should add to your list: *www.google.com/gwt/n.* It hides most graphics and lets you browse the Web much more quickly on a slow connection. In the small box near the top of the screen, type the site you want to visit.

Bookmarks

JUST AS WITH COMPUTER-BASED browsers, Silk lets you save your favorite sites as bookmarks—sites you can easily visit again without having to retype their URLs. To add a bookmark, tap the Menu button to the right of the address bar and select "Add bookmark." That's it.

To see your bookmarks, swipe or tilt to display the left panel and tap Bookmarks. You'll see a list of them, with the newest at the top and the oldest at the bottom. Tap any bookmark to visit the site.

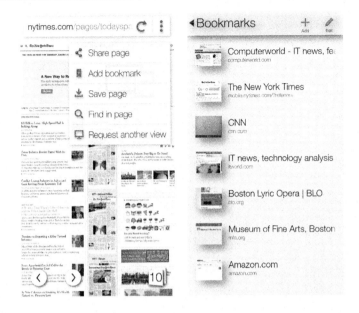

To add a new bookmark from here, tap the + sign (labeled "Add" when peeking) at the top of the screen. Type in its name and URL, and then tap OK. To remove one or more bookmarks, tap the pencil at the top of the screen (labeled "Edit" when peeking), turn on the checkboxes next to the sites you want to remove, and then tap the Delete icon at the bottom of the screen. If you decide you don't want to delete any, simply tap the big X (labeled "Cancel" when peeking) at the top right of the screen to cancel.

Not happy with the order in which your bookmarks are sorted—newest first? When you're on the Bookmarks screen, open the left panel and tap Sort at the top of the panel. You'll be able to change the order in which they're displayed, so that they're either shown in alphabetical order, by the number of times you've visited them, or by newest first.

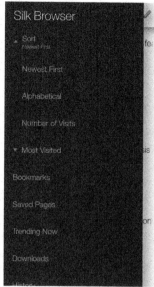

Managing Bookmarks

The Fire's Silk browser lets you do more with bookmarks as well. When you're on the Bookmarks screen, hold your finger on any and you get these options:

- **Open**. Opens the bookmarked site in the current tab.

- **Open in New Tab**. Opens the bookmarked site in a new tab.

- **Open in Background Tab**. This choice is a bit confusing. It opens the book-marked site, but rather than putting it first in your list of open tabs, it puts it last.

- **Share Link**. Tap this, and you can share the link in a number of different ways, including by email, Bluetooth, Facebook, Twitter, texting, and more.

- **Copy Link URL**. Copies the URL to the Clipboard, from where you can then paste it somewhere else, such as into an email message.

- **Edit Bookmark**. Tap this, and you can edit the bookmark's name and location.

- **Remove Bookmark**. Removes the bookmark from your Bookmarks list.

Tapping and Using Links

LINKS ON THE WEB couldn't be simpler or more convenient. Tap the link, and you get sent to a new web page. But this is the Fire phone, so there's a lot more you can do with links than just tap them. Hold your finger on a link, and a menu appears with these options:

Sometimes when you tap a link, instead of loading a web page, the Fire may take a different action. For example, if the link is to an email address, it will open the Email app, with a new message addressed to the link's email address.

- **Open.** Opens the linked page in the current window.

- **Open in new tab.** Opens the linked page in a new tab.

- **Open in Background Tab.** Opens the linked page in a new tab in the background.

- **Share Link.** Lets you share the link in a variety of ways, including via email, social networking, texting, and more

- **Copy Link URL.** Tap to copy the link's URL to the Clipboard, so you can paste it somewhere else, such as in a document or email.

- **Add Bookmark.** Adds the linked page to your bookmarks.

- **Save Image.** This option appears if you're holding your finger on an image that is also a link. It saves the image to your Downloads folder. To see the images, as well as anything else in the folder, open the left panel and select Downloads. Then tap the image you want to view.

- **View Image.** Opens the image by itself on a new page.

Saving and Viewing Web Pages

SOMETIMES YOU'LL COME ACROSS a web page that you'd like to save so you can view its information later. When you're on the page you want to save, tap the Menu button and then tap Save Page. You'll get a notice that the page is being saved, and then another notice that you've saved it. Unlike with bookmarks, when you save a page, you don't just save a link. You save a copy of the page right on your Fire phone.

To view a saved page, open the left panel and select Saved Pages. Tap any page on the list. It will look like the original page you saved, with a few small differences. First is that you'll see a small down-facing arrow to the left of the address bar, indicating that the page has been downloaded. Second is that ads are stripped out so you can focus on the important stuff. Because the page is copied to your Fire, you don't need an Internet connection to view it. By the same token, it's a snapshot in time of when you saved it, not a live web page. So if you visit the *New York Times* front page and save it one day, and then visit again and save it another day, you'll have two different copies of the front page, one for each day you saved it.

Selecting and Copying Text

AS YOU PERUSE A web page, you may come across some text that you want to take and use elsewhere, say in an email or a document. Hold your finger on the text you want to copy. When you release your finger, brackets appear around the word, and a toolbar appears with two options: Select All and Copy. Tapping Select All selects *all* the text on that web page. To copy just specific text, move the brackets until they surround all the text you want to copy. Select Copy to copy the highlighted text to the Clipboard. You get a notification that the text was copied. You can now paste it into an email, a document, and so on.

step broke a monthslong political deadlock, but it also seemed to take Iraq into uncharted territory, as Mr. Maliki gave no signal that he was willing to relinquish power.

The nomination of Haider al-Abadi, who is a me... ...Islamist Dawa Pa... ...natic late-night television appearance in which a defiant Mr. Maliki challenged the Iraqi president, Fuad Masum, and threatened legal action for not choosing him as the nominee.

As he spoke in the middle of the night, extra security forces, including special forces units loyal to Mr. Maliki, as well as tanks, locked down the fortified Green Zone of government buildings and took up positions around the city. Soldiers manned numerous checkpoints on Monday and were numerous in the Green Zone, and the atmosphere in the capital was tense.

There were no immediate signs on Monday afternoon that Mr. Maliki had taken any further steps to use military force to guarantee his survival. He was scheduled to make a public statement on television, along with other members of his Dawa Party who remain loyal to him.

Using the Menu Button

THE NOW-FAMILIAR MENU button ⁝ is familiar with good reason. As you've seen so far, it lets you add a bookmark (page 176), save the current page (page 181), and display the page in another view (page 173).

But the menu also offers two additional features. From it, you can share a web page via email, texting, Bluetooth, social networking, and more. And you can also search for a specific word or words on a web page—great if you're looking at a really long page. Tap "Find on Page," type the text you want to find, and you'll jump to it. You'll also be told how many times the text is on the page.

Go forward or backward through the search results by tapping the up or down arrow.

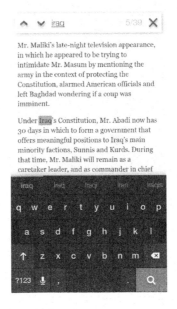

Using the Left Panel

THE LEFT PANEL IS one of the File phone's handiest innovations, giving you quick access to useful options. In Silk, it helps you manage your bookmarks, visit saved pages, and access images and downloads from the Web, as you've seen earlier in this chapter. But it has some even cooler options as well.

TIP Remember, you have two ways to reveal the left panel: swipe from the left or tilt the phone to the right.

Most Visited

This list shows you the pages you most frequently visit, with the most popular at the top. Tap any to visit it again. You'll be seeing the most recently updated version of the page. Most Visited is not a list of pages you've downloaded, but instead a list of pages you most frequently visit.

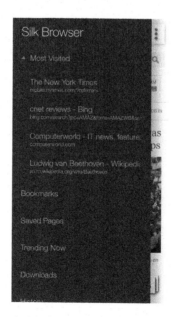

Bookmarks

Bookmarks are next on the left-panel list. For details about how to use bookmarks, flip back to page 176.

Saved Pages

Next comes Saved Pages, as discussed on page 181.

Trending Now

This automatically generated list shows you web pages that Amazon has determined are currently popular. Because the pages are taken from services and sites as diverse as Instagram, CNN, NPR, LifeBuzz, and many you've never heard about, it's a kind of X-ray of the Web's subconscious. It's visually rich—every trend is illustrated with a big picture, although you may wonder how Amazon comes up with the pictures. A trending article about fighting in Iraq was accompanied by a picture of a golfer doing what looks like a Minister-of-Silly-Walks shuffle.

Downloads

If you've downloaded anything from the Web, such as pictures and photos (page 180), here's where you'll find them.

History

Lists the sites you've visited, with the most recent ones at top. Tap the Edit button at the upper right to delete any sites from your history.

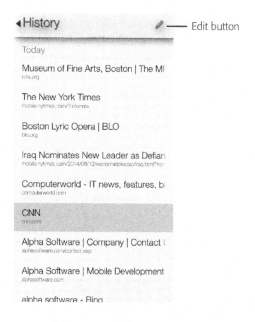

Edit button

Settings

Don't ignore this list of preferences and features for shaping your Silk browsing experience. You can change your default search engine, set how pop-ups should be blocked, turn auto-scroll on and off, and set a variety of advanced options (like Security, which is covered on page 188). Another notable setting is Silk Cloud Acceleration. This feature uses a bit of Amazon magic to speed up how quickly pages load. To turn it on and off, on the Silk Settings screen, tap Cloud. On the screen that appears, there's a setting called Cloud Features. When it's turned on, so is acceleration.

NOTE Part of the magic involves storing your browsing history on Amazon's servers. This information is encrypted and is not associated with your Amazon account. But if you don't like the idea of your information being stored on someone else's server, then turn Silk Cloud Acceleration off. For more information, see this article from the Electronic Frontier Foundation: *http://tinyurl.com/7s6883u.*

Using the Right Panel

THE SILK'S RIGHT PANEL (swipe from the right or tilt to the left) brings up a mini table of contents of the current website. Tap any listing to jump to it in the browser. Note that this table of contents is for the entire site, not for an individual page. So, for example, whether you're on the front page of *The New York Times* or reading an article, the table of contents will be the same.

Online Privacy and Security

WHETHER YOU BROWSE THE Web with a computer or with the Fire phone, there are potential security and privacy dangers out there—cookies, pop-ups, and malicious websites. So the Fire's Silk browser, just like its big-brother browsers on computers, includes the tools you need to keep you safe and protect your privacy.

Pop-Up Blocker

What's top on your list of web annoyances? Most likely at the pinnacle are pop-ups and pop-unders. They're those ugly little windows and ads that take an in-your-face stance by popping up over or under your browser so that you have no choice but to pay attention to them at some point.

Sometimes these pop-ups and pop-unders are malicious, so if you tap them they attempt to install dangerous software or send you to a harmful website. Sometimes they're merely annoying ads. Other times, though, they may actually be useful, like a pop-up that shows a seating chart when you're visiting a ticket-buying site.

So Silk takes a middle ground between blocking all pop-ups and blocking none. When there's a pop-up, it asks whether you want to allow it or not. You can, however, change that setting. On the left panel, select Settings→Pop-up Windows. Although the Ask setting gives you the most flexibility and control, you can select Always to block every single pop-up (even the good ones) or Never, to let them all through.

Cookies

Cookies are tiny bits of information that some websites store on the Fire phone for future use. When you register for a website and create a user name and password, the website can store that information in a cookie so you don't have to retype it every time. Cookies can also remember your habits and preferences when you use a website—your favorite shipping method, or what kinds of news articles you're likely to read. But not all cookies are innocuous, since they can also track your web browsing from multiple sites and potentially invade your privacy.

The browser gives you control over how you handle cookies—you can either accept them or tell the browser to reject them. Keep in mind that if you don't allow cookies, you may not be able to take advantage of cookie-based features on many sites—like remembering items in your cart on a shopping site.

To bar websites from putting cookies on your Fire phone, open the left panel and choose Settings→Your Data. Then turn Accept Cookies to Off. From now on, no cookies will be put on your phone. You can always turn this setting back on again, if it causes problems with web browsing.

Privacy Settings

If you're worried about privacy, there are a number of other settings you can adjust on the Your Data screen.

For example, you can clear your browser of data that can possibly be used to invade your privacy. Tap Clear Browser Data and then on the next screen select the kinds of data you want deleted from your phone, such as your browsing history, cookies, passwords, and so on. You also have a chance to clean out your *cache*. The cache is information the browser stores on your phone so it won't have to get that information from the Web the next time you visit that site. The cache speeds up browsing, since it's faster to grab the information—a website image, for example—from your phone than from the Web. Delete the cache to clear all that information out if you worry that it poses a privacy risk. (It's also a popular troubleshooting move.)

Remember Passwords is another setting you can turn on or off. At many websites, you log in by typing a user name and password, and other information such as your address. The browser remembers those user names, passwords, and other information, and fills them in for you automatically when you next visit. That's convenient, but it also presents a privacy risk, because someone else using your Fire can log in as you. To be on the safe side, turn this setting off (and hope that you remember all your passwords).

Similarly, you can turn off Remember Form Data. Normally, when you type information into web forms, your browser remembers it, so that it can put that information in other forms automatically. But if that worries you, turn the slider to Off.

Inbox Sent Mail

Preston Gralla   Jul 30

To: Preston Gralla

Photo I thought you'd like to see

Hi, I thought you might like to see this recent pic.

3 MB

Delete   Respond   Archive   Menu

You'll learn to:
- Set up Email
- Compose, receive, and send email
- Handle email photos and attachments
- Import and manage your contacts
- Organize your life with Calendar

CHAPTER 9

Email, Contacts, and Calendar

WHATEVER EMAIL TASK YOU HAVE, the Fire phone can handle it. Sure, it does all the usual things, such as composing, sending, and receiving mail. But it also does a great job with photos and attachments. And because email by itself is a very lonely thing, the Fire phone also has an accompanying Contacts app as well as a useful Calendar. With all three of them, you're never alone. You're connected to friends, family, and—like it or not—work.

Setting Up Email

THE FIRE PHONE HAS its own built-in Email app. You'll have to take a few minutes to set up your email account before you can use it. But fear not—as you'll see, it's a snap.

You need an existing email account to use the Fire phone's Email app, such as one from Gmail, Yahoo, or another mail provider. Once you set up the Fire phone to work with the account, your mail and contacts automatically sync with that account. In most instances, the Fire Email app will do all the configuration for you, although depending on your mail provider, you might have to do a little bit of extra work.

There's no Gmail app in the Amazon Appstore, but you can use the Fire phone's built-in Email app to access your Gmail account. The Appstore does have a Yahoo Mail app, though, so if you use Yahoo Mail, consider using that app rather than the Fire's built-in Email app.

To get started, tap the Email icon at the bottom of the screen on the Carousel, or on the Apps Grid. From the screen that appears, type your email address and then tap Next.

What happens next depends on your email provider. If you use Gmail, for example, you'll be prompted to sign into your Gmail account. For most other accounts, you'll instead go to a screen where you provide your password. Most of the time, that's all you need to do. The Fire does a fine job setting up all the behind-the-scenes techie details for you, including your incoming and outgoing mail servers.

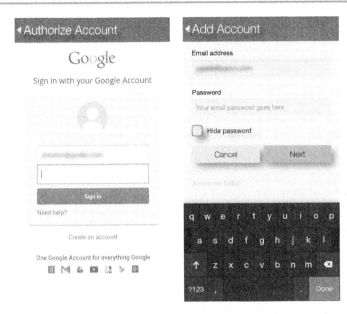

From the screen that appears, either go to your inbox or add another account.

POP3 and IMAP Accounts

There's a chance, though, that the Fire won't be able to recognize your mail provider. In that case, you'll have to go the manual route. And that means you're going to have to get a little down and dirty with some technical details. When you set things up manually, the Email app needs to know whether your account is one of three types —a POP3 account, an IMAP account, or an Exchange account.

TIP If it's a work email account, it's likely a Microsoft Exchange ActiveSync account. So you may want to check with your company's IT department for help in setting it up, or even have them set it up for you.

Here's what you need to know about each before making your choice:

- With a **POP3 (Post Office Protocol version 3)** account, the POP3 server delivers email to your inbox. From then on, the messages live on your Fire phone—or your home computer, or whichever machine you used to check email. You can't download another copy of that email, because POP3 servers let you download a message only once. If you use your account on both a computer and your phone, you must be careful to set up the account properly, as described in the box on page 197, so you won't accidentally delete email.

- With an **IMAP (Internet Message Access Protocol)** account, the server doesn't send you the mail and force you to store it on your computer or phone. Instead, it keeps all your mail on the server, so you can access the exact same mail from your Fire phone and your computer—or even from multiple devices.

- An **Exchange** account uses an email, contact, and calendar server developed by Microsoft. Generally, these accounts are used by businesses and other organizations.

To set up one of these accounts, you'll need various technical details, such as the location of your IMAP or POP3 server, and other information. If you don't have the information at hand (and face it, who does?), check with your ISP or your corporate tech support staff.

If the Fire phone doesn't recognize your account, you'll be sent to the Advanced Setup screen. Tap which kind of account you'll be setting up—POP3, IMAP, or Exchange.

Keeping Your POP3 Mailboxes in Sync

The difference between POP3 and IMAP accounts is that POP3 email lives only on whatever machine you download it to. With IMAP, a copy automatically remains on the server so you can download it again on another device. Say you read incoming email on your Fire phone, delete some of it, keep some of it, and write some new messages. Later that day, you go to your desktop computer and log into the same email account. If it's a POP3 account, you won't see those incoming messages you read on your phone, nor the ones you sent from it.

When you're using both your phone and home computer to work with the same

POPd account, how do you keep them in sync? By making your POP3 account act more like an IMAP account, so it leaves a copy of all messages on the server when you download them to your home computer. That way, you can delete messages on the phone and still see them in your inbox at home.

On your home computer, open your email program and go to your Account settings. (The location of your settings window varies depending on your email program and operating system.) Look for a setting like "Leave a copy of each message on the server," and turn it on.

You'll come to a screen that asks for technical information about your mail service. If you tap POP3 or IMAP, here's what you'll do:

- In the **POP3 server or IMAP server** text box, type in the server URL you've gotten from your provider. If your provider is named "Myprovider," there's a good chance that the POP3 server address is *pop3.myprovider.com*, and the IMAP address is *imap.myprovider.com*.

- Fill in your email address, user name, and password. The user name is typically the same as your email address.

- In the **SMTP server** text box, type in the SMTP server URL. If your provider is named "Myprovider," there's a good chance that the SMTP address is *smtp.myprovider.com*.

- Tap **Security Settings and Ports** and make sure they're accurate. You may need to check with your email provider for details.

If you're setting up an Exchange account, do this:

- In the **Exchange server** text box, type the server URL. If your company name is "Mycompany," there's a good chance that the Exchange server address is *exchange.mycompany.com*. Check with your company's IT staff to make sure.

- Fill in your user name and password. The user name is typically the same as your email address. You might want to check with your company's IT staff.

- Tap **Security Settings and Ports** and make sure they're accurate. You may need to check with your email provider for details.

Now tap "Go to Inbox" to check your mail.

Reading Email

NOW THAT YOU'VE PUT in all that work setting up your email, it's time to reap the benefits by reading your messages. Launch Email, and you'll see a list of the mail in your inbox. Mail you haven't read is boldfaced; mail you've already read is in a normal font. When you're viewing mail in a list like this, each piece of mail shows the following:

- The sender's name

- The subject line

- The date it was sent or, if it was sent today, the time it was sent

- Whether it has an attachment (you'll see a paper clip just underneath the time or date).

To open a message, tap it. Scroll up, down, and sideways in the message using the normal Fire phone gestures of dragging and flicking.

All the links you see in the email message are live—tap them, and you go to the linked web page in the browser. Tap an email address, and a new email message opens to that address. Tap a YouTube video, and the video plays.

NOTE This chapter assumes you're using your Fire phone with a Gmail account. Other accounts work in essentially the same way, although there might be some variations in the way they handle folders, labels, archiving, and so on.

The top of the email shows you the sender, his email address, the time or date the message was sent, the folder it's in, and the subject. Tap the small down arrow next to your email address, and if there's more than one recipient, it will be listed.

Down at the bottom of the screen, when you're reading mail, you'll see four icons for handling mail. Here's what they do, from left to right.

- **Delete** moves the message to the Trash folder.

- **Respond** gives you several options for responding to the message. Tap it and choose Reply, Reply All (to everyone who's gotten the email), and Forward.

- **Archive** stores the message away so it no longer appears in your inbox. But it's still available, in your email archives. (More on that on page 200.)

- **Menu** gives you a number of choices when you tap it, for moving to another folder or label (what Gmail calls folders), flagging it, marking it as unread, changing the message's label, and creating a new message. When you flag something, a small flag appears next to the message in the inbox, so you know it's important. When an email is flagged, you can tap the menu to unflag it.

Quick-and-Dirty Mail Handling

Want to delete or archive a message without reading it? Put your finger on the far right of the message and swipe to the left. (Make sure not to use the motion that you use to swipe in an entire panel.) A blue bar slides in with two icons on it, one to archive the message and one to delete it. Make your choice, or tap the X at the top of the screen to make the bar disappear. Boom—you're done.

NOTE The Fire's Email app shows messages in a threaded view, so you can easily follow entire conversations.

The exact options that appear in the bar vary according to your email account. Delete appears no matter what kind of account you have. If you're using Gmail, the other option will be Archive. If you're using an IMAP or Exchange account, it's Move. And if you're using POP3, it will be Mark Unread.

Managing Multiple Messages

Deleting, archiving, and taking other actions on one message after another can get pretty time-consuming. Managing them en masse is much quicker. To do it, when you're in the inbox, tap the Edit icon ⊠ at the top of the screen. Boxes appear next to each of your email messages. Put checkboxes next to the ones you want to act upon. Down at the bottom of the screen, tap one of the icons—Delete, Respond, Archive, or Menu. (The Email app's menu gives you options for flagging messages, marking them unread, or applying a label to them.) To call the whole thing off, tap the X button at the top right of the screen.

Number of items selected   Cancel

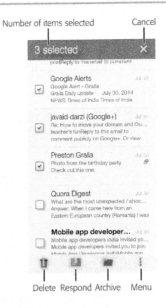

Delete   Respond   Archive   Menu

Attachments in Email

Mail lets you download attachments, including graphics like the .jpg, .png, and .gif formats and other file types like PDFs. If you get an attachment that's a picture, you'll see a thumbnail of it right in the message. Tap it, and you'll see the picture on its own screen, accompanied by several white icons. Tap the back arrow at the top of the screen to return to the email message. Tap the Download button at the bottom left to save the image so you can view it in the Photos app. And tap the Share button at the bottom right to share the image in a number of different ways, including via email, Facebook, Twitter, messaging, and Bluetooth (page 108).

The Fire phone also has a clever way for you to see all the attachments in your inbox. Swipe or tilt to display the right panel. You'll see a list of all your attachments. Tap any to display it in its own screen with the same icons shown on the previous page.

If you're downloading files other than picture files—like PDFs and other files that the Fire phone recognizes as documents—they end up in your Docs library. Otherwise, they're stored in your Email Attachments, which you can get to from the left panel in the Docs library. (For more details, see page 149.)

Writing Email Messages

WHEN YOU WANT TO create a new email message, start from the inbox and then tap the New button at the top of the screen—it's a plus sign. A new, blank message form opens, and the keyboard appears so you can start typing.

If you want a larger keyboard, rotate your phone 90 degrees.

Write your message this way:

1. **Type the recipient's address in the To field.** As you type, the mail program looks through your Contacts list, as well as the list of people you've sent email to in the past, and displays any matches. (Email matches the first few letters of first names as well as last names as you type.) If you find a match, tap it instead of typing the rest of the address. You can add as many addresses as you wish.

   If you prefer, tap the icon on the far right of the To field—it looks like a person with a + sign. You'll bring up your Contacts list. Search or scroll through it to find the person to whom you want to send the message.

2. **Send copies to other recipients.** Tap the Menu button, and then tap Add Cc/Bcc.

   Anyone whose email address you put in the Cc and Bcc boxes gets a copy of the email message. The difference is that, while everyone can see all the Cc recipients, none of the other recipients can see the email addresses you enter in the Bcc field.

3. **Type the topic of your message into the Subject field.** A clear, concise subject line is a good thing for both you and your recipient, so you can immediately see what the message is about when you scan your inbox.

4. **Type your message into the Message box.** Type (or copy and paste) as you would in any other text field.

5. **Add an attachment.** To add an attachment, tap the paper clip icon. When you do, three choices appear: "Attach a Photo," "Attach a File," and "Capture a Photo." Make your choice and attach what you want. You'll see a thumbnail of your attachment right in the message. You can add multiple attachments; tap the X button to remove one.

6. **Tap the Send icon at top right, and the message gets sent immediately.**
   To discard it, tap the X. You'll then be able to save a draft of the message or discard it entirely.

Navigation and More with the Left Panel

YOU MAY NOT REALIZE it, but the Fire phone has plenty more email features packed into it. To get to them, you have to display the left panel—swipe from the left or tilt to the right.

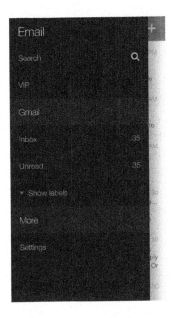

There's a lot you can do here. For a start, there's a nifty way to search through your email. Tap Search and you'll be sent to your inbox, with a search box at top. Type in the word or terms you're looking for. There's a drop-down box that lets you search for that word or term in the From, To, or Subject fields, or in the entire message. It's a great way to find what you want, fast.

There's also an odd feature on the panel called VIP. No, the Fire phone isn't complimenting you on being a very important person. Instead, it brings you to your *VIP inbox*. That inbox contains only mail from people you have designated as VIPs in your contact list (page 64).

If you use Gmail, or an IMAP or Exchange account, you should also see your mail folders (or labels, as Gmail calls them), in addition to your inbox. Tap any to go to it.

Finally, you'll see an option for settings. You'll find settings for your overall Email, Contacts, Calendar, and for any specific accounts you've set up, such as Gmail. For example, in the overall mail settings, you select whether to automatically download attachments when you're connected to WiFi, whether to include the original message in replies, whether to show embedded images, and so on. These settings apply to all your email accounts.

Your individual settings will vary from account to account. Most important are the sync and data settings, so pay special attention to them. With a Gmail account, for example, you set how often to check for new email and whether to sync your Calendar and Contacts.

Your Name
Preston Gralla

Description
Gmail

Default Account
Gmail

Sync and data settings

Sync Calendar   OFF   ON

Sync Contacts   OFF   ON

Inbox Check Frequency
Automatic (Push on Wi-Fi)

Days to Sync
One week

Signature
Append text to messages you send

Reauthorize Gmail Account

Adding a Signature

In the mail settings for each individual account, you can also automatically add a signature—your contact information, for example—at the bottom of every outgoing message. To create a signature of your own, open the left panel, tap Settings, and then tap the account for which you want to use or edit a signature. Then tap Signature. You'll see a text box with the thrilling message "Sent from my Fire"; replace that with a more creative signature of your own, and then tap OK. The signature will be appended to the bottom of all messages you send.

Working with Contacts

WHEN YOU CREATE AN email account, it syncs your contacts to your Fire so they're ready when you need them. When you want to email someone, the Mail app can grab the address right from your contacts.

Except that you might come across a slight glitch. Some email services won't sync their contacts with the Fire. If that's the case with your account, you're going to have to do things the old-fashioned way, by typing in the information yourself. It's time-consuming but not difficult. To do so, open the Contacts app by tapping its icon in the Carousel or Apps Grid.

To create a contact, tap the + button. A screen appears asking you to which account you want the contact to synchronize after you add it. You can choose from among the accounts you've already set up on your Fire. If you don't want to sync it with any, choose Device Only. You'll still be able to create the contact, but it will live only on your phone and won't sync with any other account. If you prefer, you can add a new account, and sync it with that.

NOTE You can also synchronize your Facebook friends with your Fire contacts. For details, see page 222.

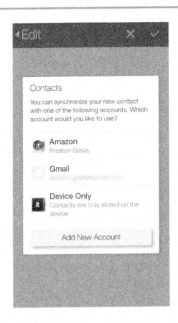

After you make your choice, you'll come to a screen with all the usual contact info—name, address, email, phone number, and so on. There are a few extra features to look for, though. You can choose specific ringtones (page 73) for phone calls and text messages associated with the contact. And if the fields you see don't cover all the bases, tap Add More Fields, and you'll get even more, including the person's nickname, birthday, and so on.

You can also include a photo of the contact. Tap "Edit image," and you get two choices. Add Photo takes you to the Photos app where you can browse for an image. Take Photo lets you use the phone's camera to take a picture of the person and include it. Just make sure to tell her to say "Cheese!"

Importing Contacts

If you're willing to do a bit of techie work, there is one more way to get contacts onto your Fire phone—import them. First you'll have to export the contacts from wherever they're stored, for example, in Outlook or another email program. Check the program's documentation for exporting contacts.

Your best bet is to export them onto a folder on your PC or Mac (ideally, in the vCard format). Then connect the Fire phone to your PC or Mac via its USB cable. When you do that, the Fire phone appears like an external hard disk to the computer, so you can transfer the contacts file onto a folder on the Fire phone (see page 141 for more details on transferring files). Then disconnect the phone and head to the Contacts app. Swipe or tilt to open the right panel, and then select Import/Export→"Import from storage," and follow the prompts.

Using Contacts

The Contacts app is a simple affair. Swipe up and down to scroll through your contacts, or search through them by tapping the Search icon at the top of the screen. When you search, you'll search every field of every contact, including not just first and last names, but also address, company, and so on. So you may get unexpected results. But that's because the app is searching fields you don't see. So let's say you're searching for a friend with the first name "Cameron." You type *Cam* and expect to see only people whose first or last name contains those letters. But you'll also see plenty of other people—people who live in Cambridge, Massachusetts, for example.

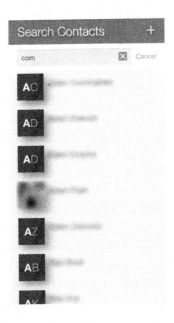

When you open a contact, you'll see all the details you've put in. Tap a phone number to call it; tap an email address to send an email; tap a street address to see the address in Maps. But there's more you can do as well. Down at the bottom of the screen are icons for deleting the contact, editing it, and sharing it (peek to reveal the labels for these icons). And there's also a Menu button at lower right that lets you perform a bit of housekeeping. If you have duplicate contacts, tap the menu icon and select Join, and you can combine two contacts into one. And if you have a contact with multiple email addresses, you can split that contact into multiple ones by tapping the button and selecting Split.

NOTE If you have a contact that does not have multiple email addresses, when you tap the Menu button, you'll instead arrive at a dialog box that lets you combine contacts using the Join feature.

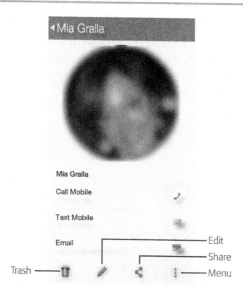

Adding a Contact to Your VIP List

Not all contacts are created equal. You probably have a few who are more important than others—your mother, for example. (At least, she'd better be.) These are the people whose contact information—and email messages—you need to get to quickly and often. That's where the Fire's VIP list comes in.

To add someone to your VIP list, simply tap the small star badge over her picture (or the generic face icon if there's no picture) in her Contact information. The badge lights up red, indicating VIP status.

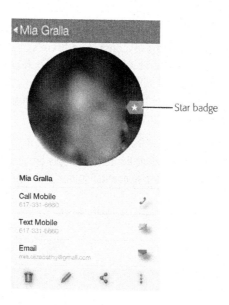

You can view your VIP list in the right panel. Tap any name to see the contact information. And when you're doing email, you can see messages from your VIPs on a special list (page 206).

To remove someone from your VIP list, go back to her contact information and tap the red badge so it turns gray again.

Filter Your Contacts

If you have contacts from multiple accounts—for example, an Amazon account and a Gmail account—they all appear in the same contact list. But you can easily narrow that down to just one account to make things simpler. Just select from your accounts in the left panel.

Contact Settings

The Contacts app doesn't have a ton of customization options, but you do get to control some things, like how often your contacts sync with their source. Open the left panel and select Settings→Contacts Settings. You can choose whether to synchronize your contacts with their source accounts (including your Facebook account). If you don't, then the contact information on your phone will stay separate and won't change when you update the information somewhere else. You can also change the order in which contacts are displayed and sorted— either by first name first or last name first.

The Calendar

THE CALENDAR, LIKE THE Email app, is not an island unto itself. You can't use the Calendar from scratch—instead, as with email, you link it to an existing Calendar, such as Google Calendar, an Outlook.com calendar, or most any other online calendar.

When you set up your email account, there's a good chance that you also told it to sync your Calendar. To make sure, swipe down from the top to get to the Quick Actions panel and tap Settings→My Accounts→"Manage email accounts" and tap your email account at the bottom of the screen. In the "Sync and data settings" area, make sure Sync Calendar is set to On. (If that doesn't work, it may be because your email provider doesn't have a compatible calendar app.)

If you don't yet have an online calendar, now is the time to sign up for one. They're free and easy to use. And as with email and contacts, they'll automatically sync with your Fire phone calendar. Once you have an online calendar, on your Fire phone, swipe down from the top to get to the Quick Actions panel and tap Settings→My Accounts→Add Account.

Viewing the Calendar

To open the Calendar, tap its icon either on the Carousel or the Apps Grid. As you'll see, it includes all the events that you have on your Google Calendar or other calendar account.

If it's the first time you've launched the calendar, it probably opens up to the day view. But there are three other views you can also use: List, Month, and Week. To switch to the List or Month view, tap the appropriate icon at the bottom of the screen.

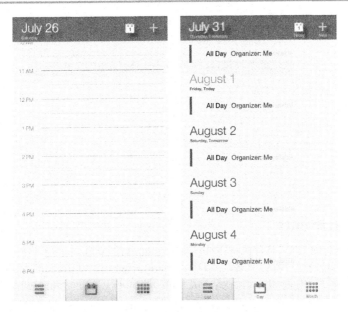

Notice something missing? Where's the Week view—there's no icon for it! When you're in any of the other three views, simply turn your phone sideways and it switches to Week view. Turn your phone back to a vertical position and it automatically switches back to whatever view you were in before.

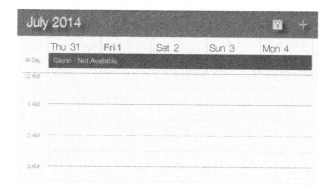

Creating Events

Creating a new event is simple. In the Calendar app, you'll see a + sign at upper right. (It's labeled "New" when you're peeking.) Tap it, and then fill in the form: a title for the event, its starting day and time, whether it's repeating, where it will be held, and so on.

If you have more than one account on your Fire phone, there's an extra step. Tap the Account field; a list of all your calendar accounts appears. Select which calendar you want this event to appear on.

NOTE Even if you've created only a single calendar account, you may see multiple accounts in the Account field. That's because Google Calendar has the ability to show multiple calendars simultaneously. So if you've set up the web-based Google Calendar to show multiple calendars, they'll also appear in your Fire calendar.

There are a few other fields that let you go beyond the basic time, date, and location. For example, the biggest advantage of having your calendar on your Fire phone is that it can beep (or buzz or vibrate) to remind you of an imminent appointment. Don't neglect the Reminder field. Your choices are many, ranging from at the start of the event, all the way out to a week before it.

But why just nag yourself about an event when you can nag others? When you include a contact or someone's email address on the event, he gets an email about it. He can accept or not, and can also see a map of where it's being held. You'll be notified whether he's accepted or turned down the invitation.

If you rely heavily on your online Calendar for organization, there's a lot you can do with the Notes field (at bottom). It's a good way to remember things about the event—for example, the names and ages of your boss's children if you're going to attend a social event with her.

When you're done, tap the checkmark at the top of the screen (labeled "Save" when peeking) to save the event to your calendar. Tap the event in the Calendar to reveal all its details. If it's an event you created yourself, you can also edit it. You can also delete the event by opening it and tapping the Trash icon.

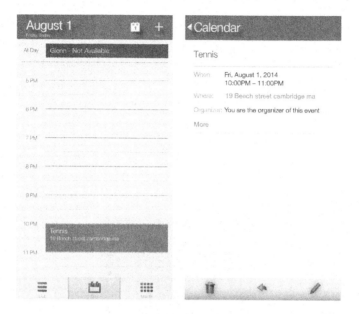

Using the Left and Right Panels

As with many other apps, the Calendar's left and right panels give you additional features. Swipe or tilt to open the right panel, and you'll see information about your next upcoming event. If it's an event you've been invited to rather than one you've created you can send a message from that screen to the event's organizer.

Swipe or tilt to open the left panel, and you'll see a list of all your calendars, with checkboxes next to those that are currently displayed. If there's one you don't want to see, uncheck it. To view it again, come back here and put a checkmark next to it.

From here, you also get to your calendar settings. Tap Settings→Calendar Settings to set global calendar settings for all your calendars, including the day you want your week to start and your time zone.

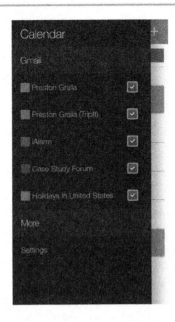

As for the right panel, it shows you the next upcoming event in your calendar.

TIP If you've linked your Facebook account to your phone (page 222), you can have your Facebook events sync to your Fire calendar. From the left panel, tap Settings→Calendar Settings and turn on Sync Facebook Events.

Hello.
Welcome to Twitter.

Sign up   Sign In

You'll learn to:

- Connect your Fire phone to Facebook and Twitter

- Sync Facebook contacts and events with your Fire phone

- Use the Facebook and Twitter apps

CHAPTER 10

Facebook and Twitter

SMARTPHONES ARE HOW MORE and more people these days keep in touch with one another. And that usually means using Facebook and Twitter.

You have two ways to use the Fire phone with those popular social networking services. You can either use the phone's built-in linking capabilities or download apps that offer even more social media savvy.

With Facebook, you can use the Fire phone's built-in linking features to sync all your Facebook friends' information and calendar events with the Fire phone's Contacts and Calendar apps. To keep up with your friends' activities, post your status, and so on, though, you must download the Facebook app.

When you link Twitter to your Fire phone in the same way, you only get a quick way to share photos and other content on Twitter. But you can download an app that lets you make full use of Twitter right on the phone.

Facebook

FACEBOOK HAS MORE DAILY users than any other social network. Millions of people around the globe use it to keep in touch with friends, play games, and do business. If you're one of them, you'll be glad to know you can sync your Facebook information with the Fire phone's own Contacts and Calendar apps. There's also an app so you can use all of Facebook's features on the go.

Syncing Facebook with Your Fire Phone

Syncing your Facebook contacts and events with your Fire phone's built-in apps takes just a few quick steps:

1. Pull down the Quick Actions bar and tap Settings→My Accounts→"Connect your social networks"→Connect Your Facebook Account.

2. On the next screen, enter your Facebook user name and password, and then tap Connect. A screen appears telling you that when you link Facebook to the Fire phone, Amazon will be able to access much of your Facebook information, including your name, public profile, your friends' email addresses and phone numbers, and more. If you don't like that idea, tap Cancel.

NOTE The text box for your Facebook user name or email address may already be filled in. Make sure it's correct before proceeding.

3. A screen appears with two options: Sync Calendar and Sync Contacts. Turn on either or both, and then tap Done.

If you ever want to unlink Facebook from your Fire phone, pull down the Quick Actions bar and tap Settings→My Accounts→Facebook. (You'll then see an Unlink button next to the word "Facebook." Tap it when you want to discontinue syncing Facebook and your phone.)

Facebook and Your Contacts

After you link Facebook and choose to sync your contacts, they're automatically imported into the Contacts app (page 108). They look like any other contacts, with no visual indication to let you know they're Facebook contacts. But you can find out: Tap the contact and then scroll down to the Sources section. If the contact has been imported from Facebook, it says so there.

NOTE You may have some contacts that you've gotten into the Fire phone from multiple places, for example, Facebook and Gmail. If so, both Facebook and Gmail are listed in the Sources section, and the contacts' information is merged.

If you only want to see your Facebook contacts, swipe or tilt to open the left panel, and on the screen that appears, tap Facebook. To see all your contacts, open the left panel again and tap All Contacts.

Facebook and Your Calendar

If you choose to sync calendar information, Facebook events will be included in the Fire phone's calendar. You'll see a small Facebook logo next to them in the day and list views, but when you tap and look at an individual event, you won't see any indication that it's from Facebook.

If you like, you can hide the Facebook events in your calendar. Open the left panel and uncheck the box next to My Facebook Events. From then on, none of them will show up in the calendar, but they'll still keep syncing behind the scenes. To display them again, turn the checkbox back on.

TIP When you link Facebook to your Fire phone, there's one more benefit: When you tap the Share button to share things like photos, you won't have to fill in your Facebook login information. Once you've linked Facebook to your Fire phone, that information is already there, making sharing easier.

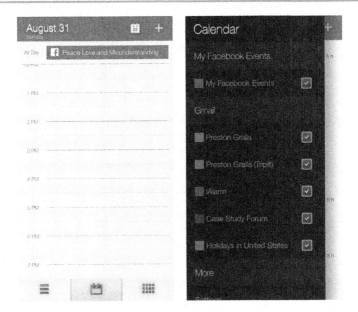

Running the Facebook App

The Facebook app doesn't come preinstalled on your Fire phone, so you have to download and install it yourself. Search for Facebook on the Amazon Appstore—either on your Fire phone or on the Web at *http://tinyurl.com/nqe5enh* and then download the app (page 259). When the download is complete, tap the Open button.

After you launch the app, sign in. If you don't yet have a Facebook account, tap "Sign up for Facebook" at the bottom of the screen and follow the directions. You'll be asked to enter a bunch of personal information to create your account.

> **NOTE** When you download the Facebook app, your phone number might already be filled in the "Email or Phone" text box. If that number is not your Facebook ID, delete that and type your normal Facebook account information.

The first screen you come to is the main Facebook page—your News Feed page. You'll see updates from all your Facebook friends, including their pictures, how long ago the update was posted...pretty much the same thing you see when you visit Facebook on the Web.

Scroll through the feed the same way you scroll through any other screen on your Fire phone: by flicking and dragging. If you've got a lot of friends, there may seem to be no end to the eternity of postings.

NOTE This chapter assumes that you already know the basics of using Facebook. If you don't, check out the Facebook website at *www.facebook.com*, click the small down arrow at the upper right, and choose Help. Select "Visit the Help Center." On the Help Center page, click the "Get Started on Facebook" link.

The Facebook app regularly checks for updates on its own. If you're like most Facebook denizens, though, you likely can't wait. To make sure that you're seeing up-to-the-minute postings, tell the Facebook app to check for any new postings. The Fire phone gives you a quick way to do that: drag the top of your News Feed down a bit and then release it. The Facebook page appears to bounce and then refreshes itself. This gesture will soon become second nature.

As you scroll down the News Feed, look at the bottom of each update. There are three buttons there: Like, Comment, and Share. Tap the appropriate button for what you want to do.

👍 Like   💬 Comment   ➦ Share

Tapping Like sends your friend a notification of your approval and adds your name to the list of people who also gave the post a thumbs-up. Tap Comment, and you'll see a list of all the comments and a box where you can add your own. Tap in the box, and the Fire phone's keyboard appears. Type your comment, tap Post, and off it goes. You can also add a picture to your comment by tapping the photo icon and choosing a photo (page 81). Tap Share, and you can share the post on your own Timeline. You get to choose who sees it—the general public, just your friends, and more.

NOTE The official Facebook app that you use may look and work a bit different from what you see pictured here. That's because the app gets updated, so you may be using a different version of the app from the one in use when this book was published.

Writing Posts, Uploading Photos, and Checking In

Now that you've got Facebook on your Fire phone, it's time to start posting. What's the point of Facebook if you can't share your innermost secrets with the world?

To post an update, tap the Status button at lower left; the keyboard appears. Type what you want, tap Post at upper right, and you'll share with the world your all-important news about your cat's changing sleep patterns.

TIP What if you've started to write a post and then decide you don't want to? There doesn't seem to be a way to cancel it. Ah, but there is. Swipe up from the bottom of the phone to dismiss the keyboard, and then swipe up again to go back to your News Feed.

But what if you want to not just talk about Kitty's sleeping patterns, but also show a picture of him sleeping as part of the post? What if you want to share the post only with specific friends? What if you want people to know the exact location of your cat while he continues with his frustrating sleep habits?

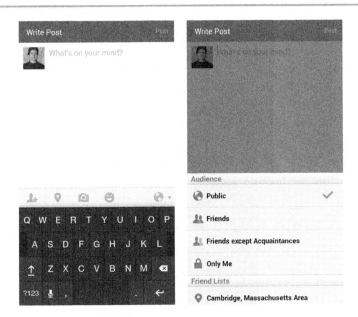

That's where the icons just above the keyboard come in. Tap the leftmost one (the picture of a person), to post to your timeline but share with only specific friends—only those you specify will be able to see it. Tap the location (Check In) button just next to it, and then choose a location. Tap the camera icon to choose a photo to include in the post. If you want to instead take a photo, then after you tap the camera icon, tap it again from the screen that appears to launch the camera, take a picture, and then embed it in your post. If you're a fan of emoticons and smiley faces, tap the icon of the smiley face to bring up a gallery of them. And tap the globe icon all the way on the right to choose how private the post will be: for anyone to see, for friends only, friends of friends, and so on.

You can also just post a photo, without tying it to a status update: back on the News Feed page, tap the photo icon in the middle-bottom part of the screen, and then choose a photo to post. And if you want to choose a location for your postings, tap the Check In icon just to the right of the camera.

Navigating Facebook's World

Look just below the very top of the Facebook app's screen—there are five icons there. They're there so that you can navigate Facebook's gigantic, ever-changing world. Here's what they do, from left to right:

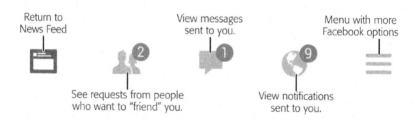

- **The mini-Facebook screen icon** brings you back to your News Feed, no matter where you are.

- **The two-person icon** shows you all the requests from people to "friend" you. Respond to any right on the screen.

- **The message balloon icon** lets you see any messages sent to you. Tap any message to respond to it, like it by tapping the thumbs-up button, and so on.

- **The globe icon** shows notifications such as likes and comments to your posts, messages from games you play, and so on.

- **The three-bar Menu icon** gives you access to countless Facebook features and settings.

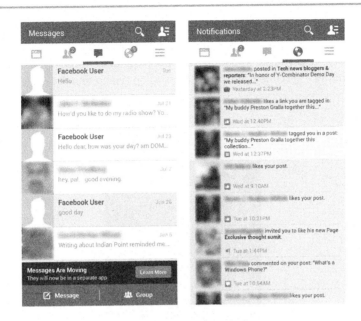

Interacting with Friends and Finding New Ones

Facebook is all about keeping in touch with friends and making new ones. To do that, you use two icons at the top right of the screen, just above the row of icons. Tap the rightmost top icon—a person with three horizontal lines—and a screen appears with a list of your friends who are currently on Facebook and available for a chat. Tap the person's listing to start the chat.

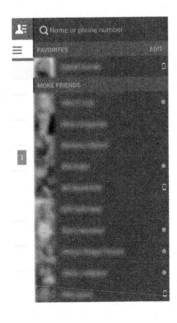

To find a friend, instead tap the Search icon, start typing a name, and you'll see a list appear. You'll see names that match the letters as you type them. So search for "ly," and you'll find people with the first names Lydia, Lynn, and Lysa, people with the last name Lyons, and so on. Keep typing to narrow the list down further.

What if you want to see a list of *all* your friends? Instead of using the Search icon, tap the rightmost one in the group of five just below it. That brings you to a menu of many Facebook features. Tap Friends from that screen to see the whole list.

Facebook makes it easy to find new friends by looking for people with whom you have friends in common. That's where the Find Friends button comes in. Tap it and a screen appears with a list of people you might know. If there's someone with whom you have a lot of friends in common, she'll show up here.

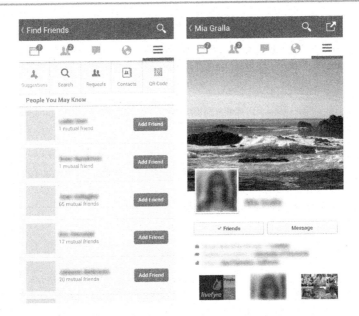

Visiting Your Friends' Timelines

Whenever you see a friend in a list, tap the person, and you'll get sent to her Timeline, where you can see her most recent updates and much more. Scroll down to see all her updates, tap the Message button to send a message, tap the Photos button to see photos, tap the Friends button to see her friends. You'll also be able to comment on any post...well, you get the idea. You can do anything with the app that you can do with Facebook on the Web.

 # Twitter

ARE YOU ONE OF the vast army of tweeters? Celebrities, politicians, and even the pope now communicate with the world in 140-character messages called *tweets*. You'll be pleased to know that you can easily tweet right from your Fire phone.

Linking Twitter with the Fire Phone

When you link Twitter to your phone, you don't get nearly the same kind of integration and syncing as when you link Facebook with it. You get one thing, and one thing alone: When you publish a tweet or share a photo, you don't have to enter your Twitter handle and password. Your information is already there.

To connect your Fire phone to your Twitter account:

1. Pull down the Quick Actions panel (page 18) and tap Settings→My Accounts→ "Connect your social networks"→Connect Your Twitter Account.

2. From the screen that appears, enter your Twitter user name and password, and then tap Connect.

NOTE The text box for your Twitter user name or email address may already be filled in. Make sure it's correct before proceeding.

3. A screen appears telling you that when you link Twitter to the Fire phone, Amazon will be able to see your tweets, whom you follow, and other information. If that disturbs you, tap Cancel to back out now.

That's all it takes. The Fire phone will now import the people you follow on Twitter. From now on, you won't have to enter your Twitter user name and password when you share via Twitter on your phone.

If you ever want to unlink Twitter from your Fire phone, pull down the Quick Actions panel as described in step 1. Tap the Unlink button next to Twitter.

Running the Twitter App

As with Facebook, you must first download and install the Twitter app (page 225). Make sure it's the official one from Twitter; Twitter will be listed as the author on the app's description page.

NOTE This section assumes you already know how to use Twitter on the Web. If you're just getting started, go to *www.twitter.com*, sign in to your account, click the gear icon at the upper right and choose Help for assistance. If you don't have a Twitter account yet, read on to set one up.

Once you install it, log into your existing Twitter account. If you're one of the remaining people on earth without a Twitter account, you can click the Sign Up button to create a new account. When you're done, you'll see your Twitter home page. Scroll through to read the latest updates from people you're following.

NOTE When you sign in, you'll be asked whether Twitter can use your location. If you're worried about your privacy, just say no.

At the top of the Home page, that navigation bar is Twitter Central for your interactions. Here's what each of the buttons do, from left to right:

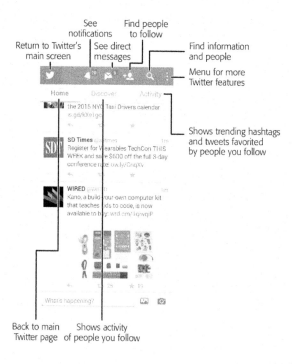

- **The bird button** takes you to Twitter's main screen.

- **The bell button shows notifications,** such as when people have retweeted or favorited tweets of yours.

> **NOTE** A retweet is when someone forwards your tweet to his own Twitter stream. Sort of like saying, "You've gotta see this tweet!" You can also retweet tweets of any of the people you follow (more on that in a moment).

- **The envelope button** shows direct messages that people have sent to you.
- **The button with a person with a + sign** helps you find people to follow and to communicate with.
- **The search button** lets you search Twitter for people and information.
- **The menu button** gives you access to other features, like working with drafts of tweets you haven't yet sent, adding a new account, and changing settings for your existing account.

Just below those buttons are three tabs:

- **The Home tab** brings you to your main Twitter page.
- **The Discover tab** shows you trending hashtags and tweets that have been favorited by people you follow.
- **The Activity tab** shows you the activity of people you follow, such as who they now follow and who they've favorited.

Creating a tweet is simple. At the bottom of Twitter's main screen is a text input box with the text "What's happening" in it. Tap the box and a new screen appears, where you type in your tweet. As you type, you'll see an indication at the top of the screen about how many characters you have left, so you can stay within Twitter's 140-character limit. Tap Tweet when you're done.

> **NOTE** If you start writing a tweet and decide to cancel it, swipe up once to make the keyboard disappear, and then swipe up again to get back to the main Twitter screen.

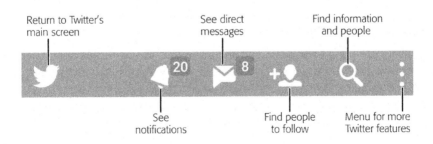

Return to Twitter's main screen

See direct messages

Find information and people

See notifications

Find people to follow

Menu for more Twitter features

But there's more to it than that. At the bottom of the screen that appears when you create a tweet are two buttons that let you enhance your tweet in the following ways:

- **The pin button** turns on location services, so your followers can see where you're tweeting from.

- **The picture button** takes you to a screen that lets you select photos you've shot on your Fire phone and send them as part of a tweet. Scroll to any picture you want and select it. Or you can snap a photo right then (page 116) and send that as part of the tweet. Just tap the camera button.

NOTE You can also include a document in your tweet. From the screen you get to when you tap the picture button, tap the icon to the right of the camera (it looks like a sun or a moon over a mountain range). Then just follow the prompts for choosing a document to send.

Taking Action on Tweets

The Twitter home screen shows you all the tweets, retweets, and direct messages from the folks you follow. You see a list of tweets, along with the person's user name and picture, as well as how long ago the tweet was made. Small icons give more information about the tweet—for example, whether it's a retweet, a direct message between people, or if the tweet has a photo attached.

Press any tweet and hold it, and a bar appears with four icons on it that let you reply to a tweet, retweet it, add it to your favorites, and share.

Respond to   Add to   Share the
a tweet   favorites Retweet   tweet

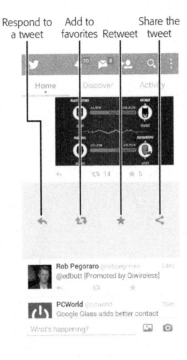

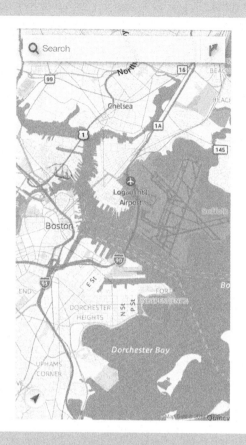

You'll learn to:
- Check out landmarks in 3D
- Find contacts, businesses, and recreational destinations
- Get directions for walking, driving, and public transportation
- Get turn-by-turn navigation

CHAPTER 11

Maps and Navigation

LIKE ANY SELF-RESPECTING SMARTPHONE, the Fire phone comes with a Maps app that shows you where you are and what's nearby. It also helps you find stores, landmarks, and restaurants and gives you turn-by-turn directions by car, foot, or public transportation. Thanks to Dynamic Perspective, the Fire phone also adds its unique twist to maps with features like 3D Look-Around and Peek.

NOTE The Maps app on the Fire phone is Amazon's own mapping app. If you're a fan of Google Maps (and its useful Street View feature), you're out of luck. As of this writing you can't get Google Maps in the Amazon Appstore. On the other hand, Amazon Maps has some tricks Google doesn't have—like Peek and 3D Look-Around.

Getting Started

BEFORE YOU GET STARTED with Maps, you need to do a few simple things to get the most out of the app. First, tell the Fire to use location-based services. That's how you tell the phone to use all its built-in features for giving Maps the most accurate information about your current location. Swipe or swivel (page 27) to pull down the Quick Actions bar, tap Settings, and then tap Location Services. Make sure Location Services is turned on. And if you want your location history stored so that Maps remembers where you've been, turn on

Enhanced Location Services (Settings→Location Services→Configure Location Based Services for your applications). Be aware, though, that if you do that, it means that Amazon will store that information about you. If you worry about your privacy, don't turn it on.

Once you set up Location Services, you're ready to use Maps. Simply type any address or point of interest in the world, and you see it on a map.

Browsing Maps

TO LAUNCH MAPS, TAP the Maps icon in the Carousel or Apps Grid. It starts out showing you a map of your current area, with your location right in the middle of it—a blue dot with a faint blue circle around it. Why the faint blue circle? As good as Maps is, it's not perfect, and so the blue circle indicates that you could actually be anywhere in the circle, if you're not precisely where the blue dot is.

TIP When you use Maps for a while, especially if you've pinched to zoom or otherwise shifted the orientation of the map, you may notice that a small compass symbol sometimes appears on the lower-right portion of your screen. What's up with that? If your map isn't oriented with north straight up, the symbol appears and shows you where north really is. To orient the map so that north is straight up, tap the compass symbol. Your map reorients itself, and the symbol disappears.

Since this is the Fire phone, you don't even need to touch the screen to look around. Just tilt it in any direction to change the perspective. Or, if you insist on using two hands, you can navigate the map by dragging or flicking. Rotate the map's orientation by putting your thumb and second or third finger on the map and twisting. Zoom in by spreading your thumb and your second or third finger or tapping twice with a finger. Zoom out by pinching your fingers, or by tapping the screen once with two fingers (the amazing two-finger tap).

As you zoom in on the map, you'll see locations of interest—museums, libraries, schools, parks, restaurants, and so on. Tap or press and hold any item, and a pin gets attached to it. You'll also see its name at the bottom of the screen, a rating if people have rated it, and a turn-by-turn icon you can tap to get directions to it.

Tap the name at the bottom of the screen, and you get a lot more details, including its address, hours, website, and phone number. All that information is live, so, for example, if it's a restaurant and you want to call in a reservation, just tap the phone number and the Fire phone places the call for you.

If you want to share the location with other people, tap the Share icon ◀, and you can share information about this place with your friends via email, social media sites, and more. Tap the Bookmark icon ▉ to mark it so you can come back to the location whenever you want. Tap the "Add to Contacts" icon ▉ to add the place to your Contacts list.

To see all your bookmarks, swipe in from the right, and the panel displays your bookmarks as well as your search history. Once you've selected a location on the map, for even more details, swipe up from the bottom, and a screen full of information about the location appears over the map, including photos and reviews from the Yelp review site (www.yelp.com). Swipe back down to return to the map.

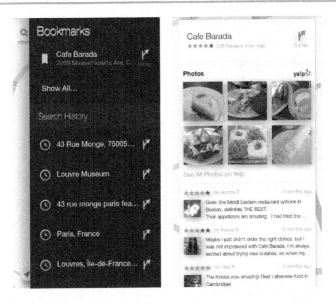

To unpin any pin you've created, select it and tap the X at the top of the screen.

Changing Your View

Just being able to zoom in and out of a map down to the street level, or all the way out to the continent level, is pretty cool, but you have still more view options, which Maps calls *layers*. A layer is a specific type of view, or information superimposed over a location. To change to a different layer, swipe from the left to display the Layers panel. You can display the Satellite layer, the Traffic layer, or both. Make your choice, and the layer changes. If you go back to the Layers panel, you'll see that your current choice has a checkmark next to it. Tap it again to turn off the layer.

The Satellite view, unsurprisingly, is an actual satellite photo of the location. Layered on top of it is the normal map, making it easy to pick out streets, street names, and landmarks. As you zoom in, the photo may at first appear blurry, and it may take a little while for the image to resolve itself, so be patient.

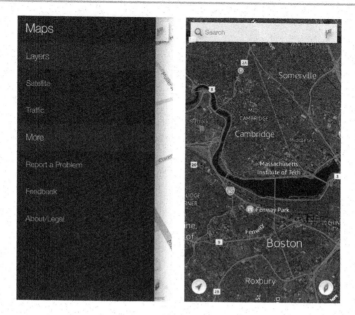

NOTE   Amazon may decide at some point to add more layers to Maps. If it does, they'll work just like Satellite and Traffic described in this book.

Traffic

The map can show you how bad the traffic is on highways and major metropolitan thoroughfares. Turn on the Traffic layer and, where available, the app indicates traffic congestion by the following color-coding:

- **Green** means the traffic is flowing nicely.

- **Yellow** indicates slower-moving traffic.

- **Red** means a traffic jam; avoid it if you can.

You can even combine the satellite and traffic layers for a rich, realistic view.

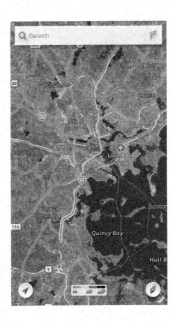

Landmarks and 3D Look-Around

MAPS ON THE FIRE phone has a unique feature that's a bit more fun than useful—displaying landmarks in cities around the world, and then letting you pan, zoom, and tilt to see them from different angles. To do it, go to a map of any city, and you'll see landmark structures or buildings in 3D. Zoom in on a landmark for best results, and then move the phone to use the 3D Look-Around feature (page 29).

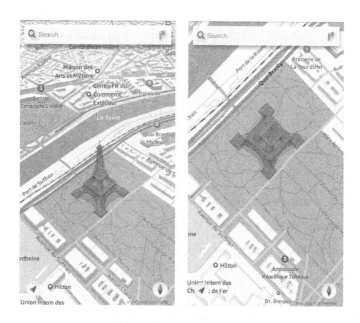

NOTE At times, you may wonder how Maps decides what qualifies as a landmark. To the dismay of Bostonians, for example, the Prudential Center is a landmark, but Fenway Park isn't. One can only hope that the next version of Maps fixes such transgressions.

How the Fire Phone Finds Your Location

MAPS' USEFULNESS REALLY COMES into play when it's combined with the Fire phone's ability to find your location. The phone finds where you are in one of three ways. The most accurate is via its built-in GPS chip. GPS works best when you have a good view of the sky. The phone also uses WiFi positioning by using information about WiFi networks near you to calculate your location. It's not as accurate as GPS, but it still works pretty well. Finally, the Fire can use cellular triangulation to calculate your location based on how close you are to nearby cellphone towers.

> **TIP** GPS and WiFi can use up a lot of juice from your battery. When you don't need them, turn them off.

Searching Maps

MAPS MAKES IT EASY for you to search for a business, address, landmark or other location. To search, tap the Search button at the top of the screen, and then type your search term into the box that appears. When Maps finds what you're looking for, it displays the location and shows a little pushpin in it. (The Fire sometimes refers to this as a *dropped pin*, particularly when you're asking for directions.)

There are countless ways to search the maps. Here are some of the most common:

- **Address**. Just type an address, including the state or Zip code. Don't bother with commas, and most of the time you can skip periods as well. You can use common abbreviations. So, if you type *157 w 57 ny ny*, you'll do a search for 157 West 57th Street, New York City, New York.

- **Intersection**. Type, for example, *massachusetts ave and cogswell cambridge ma*, and Maps displays the location at the intersection of Massachusetts Avenue and Cogswell Avenue in Cambridge, Massachusetts.

- **City**. Type, say, *san francisco ca*, and you'll see that city.

- **Zip code**. Type any Zip code, such as *02140*.

> **NOTE** As you enter search terms, Maps displays a likely list of matching results. You can speed up entering your search by choosing the right search term when it appears, rather than tapping in the entire address.

- **Point of interest or landmark**. Simply type a name instead of an address: *central park, boston common frog pond, washington monument*.

- **Airport code**. If you know the three-letter code, you can type, for example, *sfo* for the San Francisco International Airport, or *bos* for Boston's Logan International Airport.

Cambridge (02140)

- **Business**. Type *restaurants Chicago* to see a map of the city with restaurants highlighted.

When Maps finds the location, it marks it with a dot pushpin and displays the address at the bottom of the screen. Pull the address up from the bottom of the screen to get more details about it.

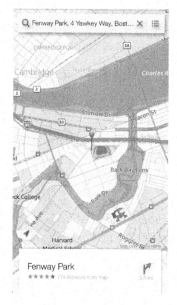

Peek on Pins

The Fire's Peek feature (page 23) works great with Maps. When you search for something, for example for restaurants in an area, you'll see dots on the map for each of your search results. To get more information about them all, move your phone slightly to the right or left to peek, and you'll see information about each from Yelp, including names and star ratings. Tap any to get more information.

Finding Contacts

MAPS CAN EASILY PINPOINT the address of a friend's home or business, as long as the address is in your Contact list. In Maps, search for the friend's name. If you have an address for the friend, it appears in the search results. Tap it, and you'll see the location on a map.

You can also do it straight from the Contacts list. Find the person in your Contacts list. Next to each of the person's addresses, you'll see a small pushpin icon. Tap the pushpin to go to the location in Maps.

Getting Directions

MAPS NOT ONLY SHOWS you locations on a map, but it can also show you exactly how to get there from wherever you are. You get to choose how you want to go: driving, using public transportation, or walking. Maps provides directions for all three, or as many as it can find.

NOTE Not all types of directions are available to all places, but you'll have more choices in major metropolitan areas.

You can get directions in many places throughout Maps, because that capability is embedded very deep in the Fire phone. For example, when you're looking at a page that gives you information about a business (page 244), there's a button you can tap to get directions.

One surefire way to get directions anywhere in Maps: Search for a location, and when you find it, tap the Directions icon  next to the location's listing at the bottom of the screen. A screen appears with a starting point, the destination, and icons for finding directions via car, public transportation, and on foot. You'll also see a list of places you've just been to or searched for.

If the Fire knows your current location, it uses that as the starting point, and shows the words "Current Location" in the starting point box. If you want a different starting point, tap Current Location and then type an address into the text box that appears.

You've already chosen your destination, but you can change it if you want. Tap End and type where you want to go. Once you've set the starting point and destination, you're ready to go. Tap which kind of directions you want, and then tap Get Directions.

Next you'll see a map with the route highlighted in blue, along with a list of various options for getting there. Each alternative route shows you the amount of time it will take, the distance in miles, and the main roads it uses. To choose one of the routes, tap it. Then tap the list icon to the right of the destination at the bottom of the screen, and you'll see a list of turn-by-turn directions.

You can scroll through the entire list of directions. To zoom in on a specific section of the route, tap it on the direction list and you'll see a close-up of the directions on the map. (For more detail on public transportation directions, see the box on the next page.)

How to Get There With Public Transportation

If you're a city dweller who gets around by hopping on buses and trains, Maps' public transportation option is your friend. When you're looking for directions, tap the bus icon, and you'll see which buses and subway lines to take. If there's more than one way to get there, you'll see all possible alternative routes. For each, you'll see the total amount of time it will take, as well as how much walking you'll need to do.

The best part: Maps taps into route schedules so that it knows which buses are scheduled to pass near where you are next, depending on the time of day.

To get more details about any of the proposed routes, tap it. You'll see not just which buses and subway lines to take, but in which direction you should be going. And if you need to do any walking, tap the icon of the person in the detailed route and you'll get sent to a map showing you where to walk.

Turn-by-Turn Navigation

TURN-BY-TURN DIRECTIONS ARE helpful, but the Fire phone offers something even more powerful—real-time turn-by-turn *navigation*, just like the GPS gizmos made by Garmin and TomTom. In Maps, type your location and final destination. Then, when you get to the screen showing you the map view of your directions, tap the Start button at top right.

NOTE   There's a difference between turn-by-turn directions and turn-by-turn navigation. In turn-by-turn directions, covered in the previous section, Maps displays your directions, but that's it. In turn-by-turn navigation, Maps talks to you to tell you where to go, and follows your progress on the map onscreen, as GPS navigators do.

Now you've turned your phone into a full-blown GPS navigator, complete with the usual robotic female voice. The Fire phone tracks your location as you drive and displays it on a map. When you're approaching a turn, it tells you what to do ahead of time. It shows you all the information you need, including distance to your next turn, current location, time to your destination, and more. So forget buying that $300 GPS unit—it's built right into the Fire phone.

NOTE For turn-by-turn navigation, you have to turn GPS on (page 241). You don't need
to turn it on to get normal directions on Maps, though.

You can also pause the turn-by-turn directions and modify your navigation
settings. Look up at the top right of the screen. You'll see the estimated time
remaining until your arrival; tap it to toggle to your estimated arrival time. Swipe
left or right (on the actual directions at the top of the screen—not on the map)
to scroll through each step of the directions. At the bottom of the screen, icons
appear:

- **Tap the X** to stop the turn-by-turn directions.

- **Tap the speaker icon** to mute the spoken directions and then turn them back
  on.

- **Tap the list icon** to see all the directions in a single list.

- **Tap Resume** to return to getting the spoken turn-by turn directions.

You'll learn to:
- Find and download apps
- Use Amazon's Appstore to download apps
- Update, manage, and uninstall apps
- Get 12 great apps

CHAPTER 12

Downloading and Using Apps

WHAT'S A SMARTPHONE WITHOUT apps? Not much but a paperweight with a phone. Apps are what give the Fire phone its amazing powers. In fact, you've been using apps throughout this book and may not even know it. When you check your email on the Fire phone, you're using an app. When you use the Calendar, Firefly, Instant Video, or Maps, those are apps.

One of the phone's great features, though, is that it lets you download and use new apps as well—and those new apps do just about everything under the sun (and sometimes things that seem beyond the range of the solar system). In this chapter, you'll find out how to get and use those apps, as well as how to uninstall and troubleshoot them. It'll also show you a few of the more amazing apps available as well.

Getting Apps from Amazon's Appstore

THERE'S ONE MAIN WAY to get apps for your Fire phone: Download them from Amazon's Appstore. Do it by tapping the Appstore icon on the Carousel or Apps Grid.

NOTE   You're not limited to the apps Amazon carries in the Appstore. You can get just about any app onto the Fire using a technique called *sideloading*. Sideloading can be tricky, but it may be worth the trouble to get an app you really want. See page 269 for details.

Before you can start downloading apps, you have to register your phone using your Amazon account. You probably did that when you activated your phone in the store or after you got it home. If not, though, when you tap the Appstore icon, you'll be asked to register. Log in using your existing Amazon account (page 288), or create a new one.

Finding apps is as easy as scrolling and browsing. When you see an app you think you might want to download, tap it. You'll come to a screen chock-full of information, including a description, screenshots, customer ratings and reviews, a list of related apps, and more.

If you worry about your privacy, scroll down to Permissions. This section tells you what kinds of things the app will be able to do if you download it; for example, access your contacts or install an app shortcut. A social media app, for example, clearly needs access to your contacts. But other apps don't—for example, a single-player game that includes no sharing has no need for information on your contacts. It could be a sneaky way to raid your phone for personal information. So read carefully, and if something raises a red flag, don't download the app.

NOTE The Amazon Appstore has far fewer apps in it than does Google Play. You'll find only apps that Amazon has approved for use on the Fire phone. As this book went to press, popular apps like Gmail, Google Maps, and even YouTube weren't available there. In fact, Amazon doesn't carry *any* Google apps for use on the Fire phone.

The Appstore has over 240,000 Fire phone apps, and you won't find them all by just browsing aimlessly. For example, you can narrow your focus by viewing apps in just a single category, like games or productivity (that is, spreadsheets and other work-related apps). Swipe or tilt to reveal the left panel, tap the category of apps you want to see, and you'll see a list of just that category of apps.

NOTE The left panel is divided into a number of sections, including the final one, More. This option doesn't give you more categories, but instead is a grab bag of features, such as getting help and contacting customer service.

Don't pay attention only to categories, though. The main Appstore screen is useful as well. You'll find featured apps and games that Amazon thinks may interest you. And it also offers a way to earn Amazon Coins—virtual currency with which you can buy certain apps and games and make in-app purchases.

Last but not least, you can search for apps. Just type what you're looking for, and you'll see screens full of what the Appstore considers matching apps. The first few matches are always on target, but after that, the results can seem random. Low down among matching results for the popular Evernote app, for example, you may see an app for turning the Fire phone into a flashlight.

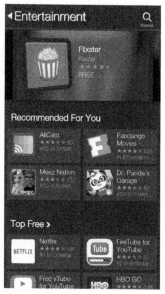

Accessing Your Apps

YOU CAN RUN APPS from the Carousel or from the Apps Grid—simply tap an app to run it. The Apps Grid has multiple screens full of apps, so scroll through it until you find the one you want. To get to the Apps Grid from wherever you are, press the Home button.

If you're like most people (including yours truly), your Apps Grid will be crammed with apps, and it might take time to scroll through the Apps Grid to find the one you want to run. Making it even more difficult is that the Apps Grid also includes your books, videos, and so on. So here's a quick shortcut to find what you want: swipe to display the left panel, and you'll see categories for apps and other content. Tap Apps to see a list of all your apps, tap Games to see a list of games, and so on. You'll then have much less to scroll through to find the app you want to run. (For details about how to reorder apps on the Apps Grid, see page 265.)

Apps in the Cloud

In the upper-left corner of the Apps Grid, you'll see two words: "Cloud" and "Device." An orange bar under one of these two words indicates which view you're currently in.

Like the name says, Device displays all the apps on your phone. The Cloud screen is less clear. When you tap Cloud, you see all the Fire phone apps you've ever acquired—some of which are on your device, and some of which aren't.

So how do the apps get onto the Cloud screen? Amazon has put some directly there, like the Prime Music app. Others are apps that you've downloaded and uninstalled from your Fire (page 267).

They're all available via the cloud, though, which means you can download them. Any apps already downloaded to your Fire phone bear a tiny checkmark. Tap any app without a checkmark to download it to your phone.

 # Fancy App Tricks

WHEN IT COMES TO running apps, the Fire has quite a few tricks up its sleeve—everything from quickly switching between running apps to creating Apps Grid collections.

Switching Between and Closing Apps

Wouldn't it be nice if you could see a list of all the apps currently running on your Fire, and switch between them? You can! Press the Home button twice, and you'll open what Amazon calls Quick Switch, which shows you all the apps you're currently running. Scroll through the list by swiping left or right until you find the app you want to switch to, and then tap it.

You might, instead, want to close the app down, because apps running in the background take up memory and can run down your battery. On the Quick Switch screen, hold your finger on the app and swipe up to close it.

When you close an app this way, you're not deleting it from your Fire. You're just closing it down.

Pin an App to the Carousel

Got a favorite app that you'd like to always have within easy reach, right in the Carousel? It's easy to *pin* an app there. The Carousel (page 13) *always* shows you apps you've pinned to it. Other apps that show up only because you've recently used them. The non-pinned apps vanish after you don't use them for a while.

To pin an app, press and hold it in the Carousel until a menu appears below it. then select "Pin to Front." If you want to remove a pinned app, select "Remove from Carousel." You can also delete it from your Fire by selecting "Remove from Device."

Reorganize Your Apps on the Apps Grid

Apps on the Apps Grid appear in random order. The only sense to it is that the last apps you've installed appear at the bottom. But what if there's an app toward the bottom that you'd like at the top, or vice versa? Fortunately, it's simple to move apps around the Apps Grid. Press and hold your finger on an icon, and then drag it to where you want it to be. When you've got it placed, release it. Voilà! Your will is done.

Create an Apps Grid Collection

One of the Fire's more useful organizational tools is the ability to group *collections* of apps on the Apps Grid. For example, you may want to group all your social media apps such as Facebook, Twitter, and Instagram into their own collection. To do it, drag an app *onto* another app, and a special screen appears. Type a name for the collection (*Social Media*, for example), and then tap OK. A collection is created with the name you gave it. You can spot a collection because its icon is made up of multiple icons—one for each of the apps you put in it. After you create the initial collection, you can drag other apps into it anytime.

Collection

NOTE When you create an app collection, the apps disappear from the Apps Grid, and appear only inside the collection. So when you name a collection, make sure the name helps you remember what's in it!

To run any of the apps in it, tap the collection to see a screen with all the apps on it. Tap any app to run it.

To remove an app from the collection, hold your finger on it and select "Remove from Collection." The app leaves the collection and goes back to the Apps Grid. When you remove all the apps from a collection, the collection itself goes away.

 You can create collections of other kinds of files, not just apps. So, for example, if you have books you want to group into a collection, drag a book onto another book, name the collection, and then drag any other books you want into it. You can't, however, mix apps and other content in the same collection.

Managing and Uninstalling Downloaded Apps

AFTER AWHILE, YOU MAY suffer from app overload: You've downloaded so many apps you don't know what to do with them. It's time to get them under control. It's easy to uninstall an app from the Apps Grid or any other screen it's on. Hold your finger on the app you want to uninstall, and then select "Remove from Device."

If you want to see more (lots more) information about any app before uninstalling it, you have to dig down into the Fire phone's settings. Swipe or
swivel to pull down the Quick Actions bar and select Applications & Parental
Controls→Manage Applications. You'll see a list of the apps you've downloaded.
Tap any, and you come to a screen full of information—the app's version number, its total size, the size of the app alone and any data associated with it, and,
toward the bottom of the screen, information about what kinds of features and
data the app uses.

You can uninstall the app—tap the Uninstall button. If you see that an app is running and you want to close it, tap Force Quit. (That button is grayed out if the
app isn't currently running.)

Managing Built-In Fire Phone Apps

THE FIRE PHONE INCLUDES plenty of built-in apps—Email, Contacts, Calendar,
Maps, Weather, Silk, Music, Photos, and more. You can't uninstall them, but
you can customize the way they work. Some apps give you more freedom than
others. Maps, for example, offers only a choice between using miles or kilome-
ters. Music, on the other hand, lets you do things like streaming or downloading music only when you're connected to a WiFi network, and preventing the
download of for-pay music.

To manage the built-in Amazon apps, pull down the Quick Actions panel and select Settings→Applications & Parental Controls→"Configure Amazon application settings." You'll see a list of all the Fire phone's built-in apps. Tap any app, and then customize to your heart's desire.

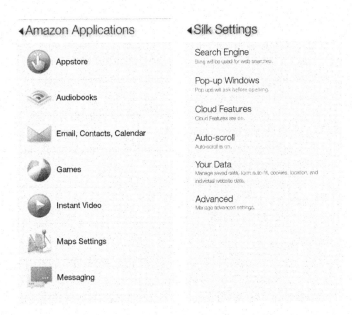

Downloading Apps from the Web

AS MENTIONED EARLIER IN this chapter, the easiest way to get apps onto your Fire phone is via Amazon's Appstore. But it's not the only way. If you're willing to live on the wild side and make a change to your Fire phone's settings, you can download them from the Web as well, using a technique commonly known as *sideloading*.

To make the settings change, pull down the Quick Actions panel and select Applications & Parental Controls→"Allow non-Amazon app installation." Next to App Installation (toward the top of the screen), tap On. You'll get a warning about installing apps outside the Appstore. If the warning doesn't scare you away, tap OK.

The Dangers of Sideloading

Be aware that you can run into problems when installing apps from the Web—sideloading—rather than getting them via the Amazon Appstore. First, the apps you find out there on the Web haven't been through the same kind of vetting procedure that they have on the Amazon Appstore. So be careful about what you download. It's a good idea to download apps only from well-known developers or well-known, trusted download libraries, such as Android Central (*www.androidcentral.com/apps*) or CNET (*http://download.cnet.com/android*).

In addition, quite a few Android apps simply don't work on the Fire. That's because Amazon changed the Android OS significantly when it built its Fire OS. For example, the Fire OS doesn't include Google Play services, so apps that require these services can't run, including Gmail, Google Drive, and so on. Some non-Google apps, such as Foursquare, require Google Play services, and they won't run, either.

Once you do that, you can download and install apps from the Web. Either visit the app developer's website to download the app, or instead head to one of the many web libraries that house thousands of apps (see the box above).

Downloading and installing from the Web takes a bit more work than getting them from the Amazon Appstore. It's a several-step process, rather than a simple all-in-one. How you download and install varies from website to website. However, the simplest way is to find the file itself using a Google search. This way, you can go directly to the developer's website and download the file straight from there. When the file is downloading, you'll see its progress on a download screen. The file will have a name like *Ghost_Commander_1.52b4.apk*. (Android apps end in the extension *.apk*.) When the file is downloaded, tap the notification.

◄Applications

App Installation
You can only install applications
from the Amazon Appstore.

OFF   ON

Collect App Usage Data
Allow Appstore to collect
information on the frequency
and duration of use of
downloaded apps.

OFF   ON

Recommendations
Product recommendations will
be shown on the Home screen.

OFF   ON

Manage Applications

NOTE Don't worry if you miss the notification after the download. You can still easily find the file. On the Silk browser's left panel, select Downloads, and you'll see the file there. Tap it to install.

◄Downloads

Ghost Commander 1.52.2.ap
/web.dl.sourceforge.net/project/ghostcommander
275 KB   8/28/2014

As when you buy from the Appstore, you'll have a chance to see what permissions the app will use (page 260). If the permissions pass muster, tap Next, and then tap Install. You'll get a notification that the app was installed. Tap Done. To run the app, head to the Apps Grid and tap the app's icon.

NOTE If you want to prevent yourself (or anyone else) from installing any more non-Amazon apps on your Fire phone, then go back to Settings and turn "Allow non-Amazon app installation" off. The apps that you installed from the Web outside the Amazon Appstore will continue to work even after you tell the Fire not to let any more apps be installed outside the Amazon Appstore.

Troubleshooting Apps

IN A PERFECT WORLD, apps would never misbehave. Unfortunately, it's not a perfect world. An app may quit the moment you launch it, or cause your Fire phone to restart, or do any number of odd things. If that happens, try these steps:

• **Launch the app again**. When you relaunch the app, it may start working properly. Whatever was making the app malfunction has gone away.

- **Uninstall and reinstall**. There may have been an oddball installation problem. So uninstall the app and then reinstall it. That sometimes fixes the problem.

- **Restart your Fire**. Just as restarting a computer sometimes fixes problems for no known reason, restarting the Fire may have the same effect. Power it down by pressing and holding the Power/Lock button, and then press and hold the Power/Lock button again to restart.

If none of this works, then it's time to uninstall the app. Don't fret; there are plenty more where it came from. Also, keep in mind that if an app refuses to shut down, you can always force it to quit. For details, see page 268.

Eleven Great Apps

THERE ARE TENS OF thousands of great apps, and a whole world of them to discover. To give you a head start, here are 11 favorites—and they're all free. You can find each of them in the Amazon Appstore.

NOTE The Amazon Appstore doesn't have nearly as large a selection of apps as the Google Play store. For example, the popular WhatsApp messaging app wasn't available as of this writing. If there's an app you want that's not available in the Appstore, you may be able to bypass the Appstore and download it from the Web. For details, see "Downloading Apps from the Web," on page 269. In this section, though, you'll only find apps you can get through the Amazon Appstore, because they've been vetted by Amazon as working properly on the Fire.

Evernote

If you suffer from information overload, here's your remedy. Evernote does a great job of capturing information from multiple sources, putting it in one location, and then letting you easily find it—whether you're using your computer, your tablet, or another Android device. Not only that, but it's free.

You organize all your information into separate notebooks and can then browse each notebook, search through it, search through all notebooks, and so on.

No matter where you capture or input information, it's available on every device on which you install Evernote. So if you grab a web page from your PC and put it into a notebook, that information is available on your Fire phone, and vice versa. You can capture information from the Web, by taking photos, by speaking, or by typing or pasting into existing documents.

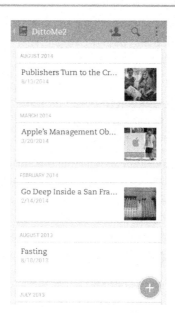

The upshot of all this? Evernote is the best app you'll find for capturing information and making sense of it all.

Vine

This Twitter-created app does for videos what Instagram does for photos— it lets you create simple, 6-second looping videos, and then post them and share them with others. And it lets you check out looping videos that others create as well.

Creating a looping video is simple. Launch the app and tell it to create a video; it switches to your camera. Hold your finger on the screen to record; lift your finger from the screen to stop recording. A bar across the top shows you how much you've already recorded of your 6 seconds of fame, and how many seconds are left. After you've created it, it's ready to go and can be shared via Twitter and Facebook. There's also a Vine feed, so that people can subscribe to your feed, and you to theirs, so you can see a constantly changing feed of these little videos.

Instagram

This popular app combines excellent photo filtering tools with photo sharing. It's a great way to play with photos and share them with friends and family, especially via Facebook. And it's also great for seeing photos that others take, and not just friends and family, but celebrities as well, many of whom seem to be Instagram junkies.

Load Instagram, use its tools to snap a photo, and then use its magic. You can add a frame and select a specific focus point, but that's just the beginning. It's the filters that you'll care about, well over a dozen. Retro, black and white, Nashville, Hefe, Kelvin...you'll find anything you need to indulge your inner photographer.

After you're done, it's time to share, and Instagram makes it easy. You can share via Twitter, Facebook, Flickr, Tumblr, and Foursquare without any heavy lifting, because they're all built into the app. And you can use the app to follow others' photos as well.

IMDb

Amazon-owned IMDb claims to have the world's largest collection of information about movies and TV shows, and it may well be right, with more than 2 million movie and TV titles, and more than 4 million actors, actresses, directors, and others. If there's any piece of information you want about a movie or a TV show, IMDb probably has it.

IMDb's information is also integrated directly into the Fire phone's guts. So, for example, when you're watching a TV show or movie on Amazon Instant Video, you'll be able to get information about it from IMDb as you're watching.

iHeartRadio

This excellent music-streaming app lets you create your own radio stations based on your favorite artists. But it goes beyond that and also lets you listen to any of thousands of radio stations that stream on the Web.

Fire owners will particularly enjoy it, because it works with Firefly. Use Firefly to identify a song, and then iHeartRadio can build a radio station based on it. (For details, see page 51.)

StubHub

This app offers an easy way to buy and sell tickets to sports and entertainment events. You also get venue information, including maps to make it easier to buy the right ticket. The app also makes use of the Fire phone's Dynamic Perspective feature—when choosing seats, tilt the phone to see the view you'll get from your seat in multiple directions, just as if you were there. StubHub also uses the Fire's Firefly capabilities by recognizing a song, finding out if the band is playing nearby, and then letting you buy tickets.

Saber's Edge

Playing games on the Fire is a whole different experience than on the average smartphone. Dynamic Perspective lets you peek around corners and control games by merely tilting the phone. So far, not many games take advantage of Dynamic Perspective, but Saber's Edge does, and does a great job of it—no surprise, given that Amazon developed it.

It's a combination fighting and tile-matching game where you need to match tiles across all the sides of a rotating cube. Use Dynamic Perspective to peek around the cube's corners or to get to the next face of the cube.

Snow Spin

Here's another game that makes use of Dynamic Perspective. You're a snowboarder, and you've got to smash ice blocks to set penguins free. Tilt the phone to move in either direction, tap it to jump, and just plain have fun.

Dynamic Perspective and Games

Dynamic Perspective adds a lot to games like Saber's Edge and Snow Spin. But even months after the release of the Fire, games that use Dynamic Perspective remain few and far between. Developers must recode their games to take advantage of it, and so far, they're not rushing to do that work.

The most popular Android games don't use Dynamic Perspective, including Candy Crush Saga, the various Angry Birds games, and many others. The best way to find games that use Dynamic Perspective is to search for "Dynamic Perspective" in the Amazon Appstore. When you do, though, you'll find only a handful of games.

Fitbit

This app works with the popular line of Fitbit activity trackers and smart scales. The trackers and scales track the number of steps you take, the distance you go, stairs climbed, the calories you've burned, your sleep habits, weight, body fat percentage, and more. The app lets you see the stats in real time, helps you set and track fitness goals, track your eating habits with a food database, and more.

MyFitnessPal

If you're looking to gain weight, lose weight, or maintain your weight (which pretty much covers us all, doesn't it?) you'll want to give this free app a try. Use it as a diary of everything you eat, and it tracks your calories and helps you keep on track to your goal. It's exceptionally easy to track what you eat, because it uses Firefly to recognize food from packaging and bar codes, and then automatically grab nutritional information from a database.

TripIt

Anyone who travels needs this app. It's a companion to the TripIt website that makes it easy to track and organize all your travel, and share it with others. When you get email confirmations for plane tickets, hotels, and so on, and forward them to TripIt, the service organizes it all into itineraries for you. You see your trips at a glance in this app or on the Web, and can easily clue other people in to your travel plans.

Wallet

The Wallet app is, at this writing, still in beta (which means it's still being tested), but you may already have it on your Fire phone—check your Apps Grid. This app lets you organize and use all your gift and loyalty cards without lugging around (and possibly losing) the physical cards. To add a card to this virtual wallet, you can either type its name and number or scan its barcode. Tapping "Scan barcode" turns on the Firefly camera (page 41); center it over the barcode. You can also store images by taking pictures of the front and back of the card, so you can quickly identify the cards in your list. When you're ready to shop, tap the card in your list. The card's barcode will appear on the screen so the clerk can scan it.

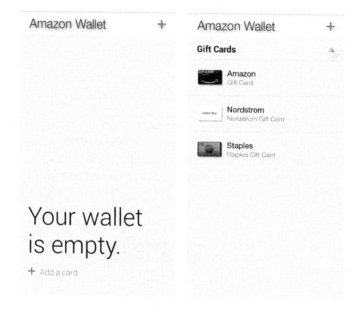

Appendixes

◀ **System Updates**

Current Version
Fire OS 3.5.1 (111009920)

Installed Wednesday, August 13, 2014 3:18 PM
EDT

Update Status
Your software is up to date
Last checked on Saturday, September 13, 2014
9:09 AM EDT

Check Now

You'll learn to:

- Select a service plan
- Set up the Fire
- Make service plan changes on the Web
- Upgrade the Fire's software

Setup and Signup

SETTING UP YOUR FIRE phone is easy, especially if you buy it at your wireless carrier's store. There, the sales folk will walk you through the process of activating your phone and signing up for a plan. If you buy your Fire phone over the Web, you set everything up either on the Web or over the phone. This appendix tells you everything you need to know.

> **NOTE** At this writing, the only carrier available for the Fire phone is AT&T. But others may come on board, so check with your carrier.

Choosing a Plan

WHEN YOU BUY A FIRE, you'll usually get it in conjunction with a one- or two-year service plan in addition to the cost of the phone. The cost of the plan may vary according to how many minutes of talk you want each month, and whether you want text messaging as well. One very good deal is signing up for a two-year contract with AT&T and getting the Fire for 99 cents. (For more information about contracts, see page xvi.)

You'll have to buy a data plan, and you may also have a data cap, which means that if you use more than a certain amount of data in any given month, you'll pay extra. You can buy texting on a per-text or unlimited basis. Heavy texters will find the unlimited plan cheaper in the long run, while occasional texters will do better paying on a per-text basis.

Your Phone Number

The phone number you use on your Fire depends on whether you already have an account with your existing provider:

- **Keeping your old number**. If you already have an account with your provider, you can have an old cellphone number transferred to your new Fire phone. Transferring the number to your new phone usually takes an hour or less. During that transition time, you can make calls with your Fire phone, but you can't receive them.

- **Getting a new number**. If you don't already have an account with your provider, the company will assign you a new phone number. It'll try to give you one within your area code, and it may have several numbers you can choose from. Once you get the new phone number, you can start making and receiving calls.

TIP If you already have an account with your provider, you may not be able to get the reduced price when you switch to a Fire phone. Providers usually require you to have your current phone for a certain amount of time—usually a year or more—before you can get a reduced price for buying a new phone. However, if you have a family phone plan, there may be a workaround. If one of your family members' lines is eligible, you may be able to get the reduced price. Just make sure that your provider connects the Fire to your phone number and not the family member's.

Using Your Amazon Account

YOU NEED AN AMAZON account to use the Fire phone. When you first use the phone, a screen appears asking you to register your phone. Tap Register, and then fill in your Amazon account information (email and password), and you'll be set to go. If you don't have an Amazon account, don't worry, because you can set it up right from your phone. After you tap Register, tap Start Here and follow the directions for setting up an Amazon account.

Making Account Changes on the Web

YOU CAN CHANGE THE details of your plan anytime—for example, adding new services, or taking away old ones—via the Web. Sure, you can do the same thing by showing up at one of your carrier's stores, but it's usually much easier to simply select services on the website.

Upgrading to the Newest Software

AMAZON'S FIRE OS IS the Fire phone's operating system. The Fire OS is built on top of Google's Android, so it's a bit of a mashup. For example, the Apps Grid looks a lot like the apps screen on a normal Android phone, but the Carousel is unique to Amazon devices like the Fire phone and Kindle Fire. (For more detail, see the box on page 20.)

Amazon regularly upgrades the Fire OS, but you don't need to buy the upgraded software, or even download it. Instead, it comes automatically to your phone, by an over-the-air (OTA) update. If an update has arrived, you'll get a notification.

There's no need to put all your trust in automatic updates, though. To check whether your phone has the latest and greatest software from Amazon, swipe or swivel (page 27) to pull down the Quick Actions panel and tap Settings→Device→Install System Updates. The System Update screen will let you know whether your system is up to date. If an upgrade is available, the phone will ask if you want to install the new software, and then do so by OTA update. To check whether one is available, tap Check Now.

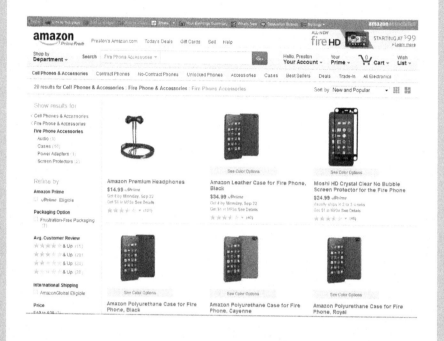

You'll learn to:
- Buy useful accessories
- Find the best places to shop

Accessories

YOU CAN FIND PLENTY of accessories to help get more out of your Fire phone. You can get products to protect its case and screen, connect it to a car charger, and more. In this appendix, you'll get the rundown on what types of accessories are available, and a sampling of where to buy them.

Useful Accessories

IF YOU ENJOY TRICKING out your car or accessorizing your outfits to the max, you can find all kinds of Fire phone bling—like decorative cases, colorful ear-buds, and even dangling charms. If, however, you're in the market for something useful, consider the following:

- **Cases**. Cases protect the Fire phone against damage from drops and spills. You choose from a wide variety of case types, depending on your style preferences, budget, and needs. You'll find hard protective cases, rubberized cases, holsters with belt clips, and more.

- **Screen protectors**. These thin sheets of plastic safeguard your phone's glass screen, greatly reducing the risk of scratches. They're thin enough so that you won't notice they're there.

- **Car chargers**. Plug one end into your Fire phone and the other into your car's power outlet, and you can charge your phone while you're on the go. Your car may have a two-prong outlet or an older-model cigarette lighter. You can get chargers for either.

- **HDMI adapter**. This adapter lets you connect your Fire phone to a TV and watch HD video from your phone on the TV's big screen.

- **Chargers and cables**. There are plenty of battery chargers and USB cables you can buy to supplement or replace the ones that came with your Fire phone, including portable chargers.

- **Bluetooth headset**. With one of these, you can talk on your Fire phone by speaking into the wireless headset (page 74).

- **Headphones**. You'll want these to listen to your music collection. Headphones can be as cheap as $30 or less for basic ones without great sound, or up to $300 or more for high-end noise-canceling ones. It's a good idea to try them out—or at least read reviews—before buying.

- **External and Bluetooth speakers**. Want to share your music with others? Get external speakers to plug into the Fire. There are plenty made for portability, with surprisingly good sound. Increasingly popular are Bluetooth speakers, so that you don't need to physically connect your Fire phone to them. (See page 74 for info on making Bluetooth connections.)

Places to Shop

THERE ARE COUNTLESS PLACES online where you can buy Fire phone accessories, but you want to make sure to order from someplace reputable, and where they know what works with your phone. Here are a few of the best:

- **Amazon (*http://amzn.to/1FSZRDj*)**. No great surprise, but Amazon is a great place to find Fire accessories. To avoid typing this URL, just go to *www.amazon.com* and, at the top of the page, choose Fire & Kindle→Fire Phone→Accessories. Or search for "Fire phone accessories" in the search box on any Amazon page.

- **Shopandroid (*http://www.shopandroid.com/*)**. Sells accessories for many types of Android phones, including the Fire phone.

- **Best Buy (*www.bestbuy.com*)**. Both the physical stores and the website are well worth checking out for a wide range of products. You can even order online and choose a nearby store where you'll pick up what you're ordering, and it'll be waiting for you when you get there.

- **Your carrier**. The place that sells the Fire phone also sells accessories. Go to your carrier's website and search for accessories. There may be some available at its brick-and-mortar stores as well.

You'll learn to:

- Make sure your software is up to date

- Fix a frozen phone

- Troubleshoot email settings

- Reset your phone

- Find where to go for free help

Troubleshooting and Maintenance

THE FIRE IS VULNERABLE to the same kinds of problems that can occur in computer operating systems. Like any electronic device, the Fire phone can be temperamental at times. This appendix gives you the steps to follow when your phone is having issues.

Make Sure Your Software Is Up to Date

NO COMPUTER OR PHONE is ever perfect; neither is any operating system. So phone makers and software companies constantly track down and fix bugs. They then send those fixes to you via software updates delivered wirelessly—called over-the-air (OTA) updates (page 289).

So if you have a bug or other nagging problem with your phone, there may already be a fix for it via one of these updates. You shouldn't have to do anything to install these updates, because they're delivered to you automatically. On the off chance that you didn't get your update, you can check and download it manually. Pull down the Quick Actions panel and tap Settings→Device→Install System Updates. The phone will let you know whether your system is up to date. If an upgrade is available, the phone will ask if you want to install the new software. To check if an upgrade is available, tap the Check Now button.

Fixing a Frozen Phone

IT'S EVERY PHONE OWNER'S nightmare: Your phone won't respond to any of your taps. There's seemingly nothing you can do. Often, your best bet is to simply restart the phone. Hold down the power button, and from the screen that appears, select Restart. In many cases, this thaws your frozen phone.

Resetting the Fire

IF ALL ELSE FAILS, you may need to reset your Fire phone—that is, delete all its data and return it to the state it was in before you bought it, with all the factory settings replacing your own. Your contacts, social networking accounts, email and Gmail accounts, and so on all get deleted, so save this step for a last resort.

Pull down the Quick Actions panel, and then select Settings→Device→Factory Reset Your Fire. Make sure that the box next to Back Up Device Before Reset is checked. That means your settings and other application data are backed up to Amazon's servers, and after you reset and log back in, the Fire phone will automatically restore the data and settings. To perform a reset, tap Reset. That erases all the data on your phone. (Now you see why it was so important to back up first.)

Warranty and Repair

THE FIRE PHONE COMES with a one-year warranty from Amazon. If you bought your Fire phone from someone else, or if someone gave it to you, the warranty doesn't transfer to you; it covers only the person who originally bought the phone. The usual types of caveats apply to the warranty—if you've misused the phone, dropped it into water, and so on, the warranty gets voided.

For more details about your warranty, read the warranty guide that came with your phone.

Where to Go for Help

IF YOU NEED HELP with anything on the Fire phone, there's nothing better than live help from Mayday—see page 36.

If you're the DIY type, the best place to go is to Amazon's "Support for Fire Phone" page. It has plenty of helpful information and a searchable database of help. It's well worth the visit. Head to *http://amzn.to/1FTpYtX*.

Also useful is Amazon's Fire Phone User's Guide, a downloadable PDF. Get it here: *http://amzn.to/1FTq6tA*. If that URL no longer works, do a web search for "Amazon Fire Phone User's Guide."

Settings ▼

Search Settings 🔍

Wi-Fi & Networks

Display

Sounds & Notifications

Applications & Parental Controls

Battery & Storage

Location Services

Lock Screen

Keyboard

Phone

My Accounts

Device

You'll learn to:

- Configure WiFi and wireless
- Customize the Fire's display
- Manage applications and parental controls
- Turn location services on and off

Settings

ONCE YOU SIGN INTO your Amazon account on the Fire, it's set up for you and ready to go. But what if you want to turn on Airplane mode, customize the screen, or configure the way emergency alerts work? You turn to this chapter, which describes all the phone's settings and explains what they do for you. To get to the Settings screen, swipe or swivel to pull down the Quick Actions panel from the top of the screen and tap Settings—it looks like a gear. You can also tap Settings on the Apps Grid or on the Carousel.

To use the Settings screen, browse until you find the heading of the category whose settings you want to adjust, such as Keyboard or Lock screen. Tap the heading to expand it to see options, tap it again to see only the headings, without options. Or use the disclosure triangle at the upper right. Tap to change it from pointing down (labeled "Expand All" when peeking) to up (labeled "Collapse All" when peeking).

NOTE Some features and customizations found in Settings are covered in more detail elsewhere in this book. For example, to learn the basics of connecting to WiFi networks and managing those connections, turn to Chapter 7 "Getting Online: WiFi, 4G, and Mobile Hotspots."

Wi-Fi & Networks

HERE'S WHERE TO GO for all things WiFi and other wireless settings. It's also where you check your cellular data usage.

Connect to Wi-Fi

As detailed in Chapter 7, this section lets you connect to WiFi networks and manage them. Turn to page 154 for details.

Enable Airplane Mode

Tap this setting, and you get a screen that does more than just turn Airplane mode on and off. It gives you options to turn Cellular Data on or off, turn Data Roaming on or off, see Data Usage, and other advanced settings (page 162). Turning on Airplane mode turns *off* all of your radios, such as WiFi, Bluetooth, and your cellular connection.

Pair Bluetooth Devices

Turn this on and then tap "Pair a Bluetooth Device" to find nearby devices—computers, other cellphones, Bluetooth earpieces, and so on. The pairing process will vary depending on the device, so check for instructions. When you tap "Pair a Bluetooth Device," a screen appears listing all detected devices. If a Bluetooth device isn't showing up, tap Scan to try again. If it still doesn't appear, check to make sure the other device has Bluetooth visibility turned on. You can also try turning Bluetooth off and then on again.

Set Up a Wi-Fi Hotspot

This lets you turn your phone into a WiFi hotspot to give Internet access to other devices. For details about how it works, see page 159.

Enable NFC

Turns on a radio technology called NFC (near-field communications). NFC can do a variety of things, such as let you transfer files and information between devices by touching them together, grabbing information from NFC tags embedded in public locations. Unless you plan to use NFC, keep this setting turned off to save battery life.

> **NOTE** NFC is also used for mobile payment, although at this writing, apps for the Fire phone are few and far between. One example is an app released in July 2014 called Pay by Phone, which lets you pay at certain parking meters by tapping your Fire phone.

Turn Off Cellular Data Access

Tap this and you get to a screen with many options, including turning off access to cellular data, putting your Fire phone into Airplane mode (page 162), and turning data roaming on or off. With data roaming, you get access to a cellular data network even if you're not within range of your providers. Be careful with this feature, because, while it's useful, you could also end up spending a lot of money when you connect via data roaming. Check with your service provider to see if you'll be charged.

Data usage

See the next section for details about what this setting does.

See Your Cellular Data Usage

If you have a data plan in which you pay for data use over a certain limit, tap Data Usage after you tap "Turn off cellular data access." You use data when you browse the Web, send and receive email, and so on. That screen shows you your data usage by app such as Maps, Email, and Facebook, and the total amount of data you've used. It sorts by putting the biggest data users on top. More important is the Manage Data Limit entry. Tap it and you can have your Fire phone send you an alert when you're getting near your data limit. You can even have your phone automatically turn off all data use when you reach your limit.

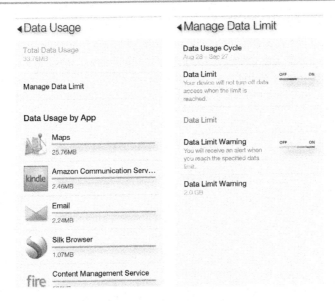

◄Data Usage

Total Data Usage
33.76MB

Manage Data Limit

Data Usage by App

Maps
25.76MB

Amazon Communication Serv...
2.46MB

Email
2.24MB

Silk Browser
1.07MB

Content Management Service

◄Manage Data Limit

Data Usage Cycle
Aug 28 - Sep 27

Data Limit   OFF   ON
Your device will not turn off data
access when the limit is
reached.

Data Limit

Data Limit Warning   OFF   ON
You will receive an alert when
you reach the specified data
limit.

Data Limit Warning
2.0 GB

Change Your Mobile Network Operator

This setting does what it says—lets you change your mobile service provider.

Display

HERE'S WHERE TO GO to manage all things having to do with the phone's display.

See your cellular data usage

Change your mobile network operator

Display

Adjust screen brightness

Turn off automatic screen rotation

Show status bar

Change time to sleep

Share your screen via Miracast

Configure low motion settings

Sounds & Notifications

Applications & Parental Controls

Battery & Storage

Location Services

Adjust Screen Brightness

This section gives you settings for how bright you want your screen to be. The top two options on the screen are Auto Brightness, which lets the Fire determine the optimum brightness for the screen, and Display Brightness, which lets you set the brightness manually. It's a bar—swipe left for dimmer and right for brighter.

NOTE When you tap "Adjust screen brightness," "Turn off automatic screen rotation," "Show status bar," or "Status Bar" you're sent to one single screen—shown above—that has options for all of them.

Turn Off Automatic Screen Rotation

This setting prevents the Fire phone from reorienting the screen between portrait and landscape modes when you rotate it.

Show Status Bar

Out of the box, the Fire phone displays the Status bar (page 9) all the time. Confusingly, this setting is set to Off, meaning that the Status bar is displayed all the time. Turning it *On* tells the Fire phone to show the Status bar *only* when you peek on the Home or Lock screens, or swipe down from the top of the screen. Why Amazon didn't call this setting "Hide Status Bar" is anybody's guess.

Even if this setting is turned on, the Status bar won't appear if you're using an app that displays full screen—like the video player.

Change Time to Sleep

This setting lets you set how long it takes for the Fire to go to sleep after a certain amount of inactivity. You can select anywhere between 15 seconds and never. It lets you set the "Screen Timeout"—that's the heading shown on the option screen when you tap "Change time to sleep."

Share Your Screen via Miracast

Miracast lets you display whatever is on your Fire phone's screen on the screen of *another* device, such as a television. In order for this to work, the TV or other device must support Miracast (and be turned on). Check the device's documentation for details.

Configure Low Motion Settings

This setting lets you put your Fire phone into (or out of) Low-Motion mode, in which features such as Dynamic Perspective are disabled. For details, see page 30.

Sounds & Notifications

THIS SECTION LETS YOU adjust everything to do with the Fire phone's sounds and notifications—for example, your ringtones, and how the Fire phone alerts you when you have an incoming notification.

Change Your Ringtone

Here's where you change your ringtone. After you tap this setting, tap Sound, select a new ringtone, and then tap its checkbox. Underneath Sound, there's also a Vibrate setting that lets you determine whether your phone should vibrate for incoming calls when you've set the ringer to on.

Manage Notifications

This setting lets you select specific notification sounds that each app uses. So you can have one sound for Email, another for Facebook, and so on.

Select Ringtones for Specific People

Tap this option, and a screen appears with all your contacts. Tap the person for whom you want to select a specific ringtone, and then select a ringtone from the screen that appears and tap the checkmark. That ringtone will then sound when the person calls.

Select Text Message Tones for Specific People

This option works just like the previous one. Tap it, and a screen appears with all your contacts. Tap the person for whom you want to select a specific notification tone for text messages, and then select a tone from the screen that appears and tap the checkmark. That one will then sound when the person sends a text message.

Change Volume Levels

This screen lets you set different volume levels for different media, including music, video, and games; the ringer; voice; and the alarm.

Change Touch Feedback Settings

The Fire phone can be as noisy or as quiet as you like. In this section, you can choose whether to have the phone make sounds when you use touch input, have the Fire vibrate in response to your touch, and have the keyboard or keypad sound tones when you use them.

Applications & Parental Controls

HERE'S WHERE TO CONFIGURE overall settings for applications, as well as for the Fire phone's parental control features.

Configure Amazon Applications Settings

This setting lets you change how the Fire phone's built-in Amazon applications work. You can change settings for Maps, Messaging, Music, Photos, Shop Amazon, Silk, and Weather. From the screen that appears, tap the app whose settings you want to change. What you'll see will vary according to the app. For example, when you tap the video app, you can adjust plenty of settings, including whether you should be asked before downloading HD videos, and the quality of the audio accompanying the video.

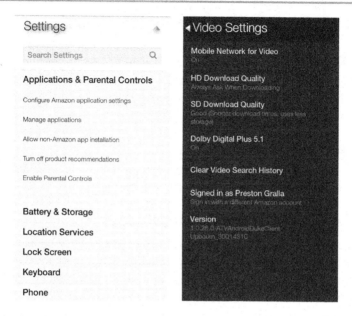

Manage Applications

This lets you manage apps, and lets you control things such as forcing an app to quit, uninstalling it, and more. For details see page 267.

Allow Non-Amazon App Installation

Out of the box, the Fire phone lets you install only apps you download from the Amazon Appstore. However, this setting lets you download other apps as well. To learn how to do it, see page 269.

Turn Off Product Recommendations

Normally, your Fire phone recommends products for you on the Home screen when you're in the Carousel (see page 13). You can get rid of them by setting Recommendations to Off. This screen also has an option called "Collect App Usage Data," which lets Amazon see which apps you're using and how often. Developers use this information to improve their apps, and it's all done anonymously, but turn this setting off if you'd just as soon not share details of your app usage.

Enable Parental Controls

The Fire phone lets you prevent children (or adults) from using certain features and apps. To activate these features, you must enter a password.

After you turn on parental controls, a screen appears where you can create a password and choose which features to block. For example, you can block web access but allow access to the camera, and so on.

Here's what you can restrict access to:

- Web browsing
- Email and calendars
- Social networks
- Camera
- Making purchases from Amazon, including the Appstore
- Playing movies and TV shows from Amazon Instant Video
- Specific types of content (such as music, video, books, and apps)
- Wireless and mobile network connectivity

In some instances—for the web browser, email, calendars, social sharing, and camera—you only have two choices: either block or allow access. In other instances, you can allow access but only by password. For example, you can require a password for making purchases, playing video, using WiFi, using location services, and accessing the mobile network.

Battery & Storage

BATTERY AND STORAGE ARE two different things, but the Fire phone puts the settings for both in one category. In both cases, the settings help you control how the phone uses these valuable resources.

View Battery Usage

This item shows your battery level and also displays how much of your battery is used by the screen, operating system, services, and apps. Tap any item, and you can either configure how the item uses power, or view details about its power use. For example, when you touch Screen, you can reduce screen brightness to save power.

View Available Storage

Shows how much memory is in use, and how much is still available. Better yet, it shows how much storage is used by apps, games, books, and so on. They're sorted by category, with apps taking up the most space on the top. Tap any category, and you'll come to a screen that lists all the content in it—for example, all your books. You can then select individual books to delete.

NOTE When you delete books and magazines, they're deleted from your Fire phone but still live in your Amazon account in the cloud. So you can always download them again. For details, see page 148.

You can also free up space by archiving old items you haven't used in a while by tapping Clear Storage Space.

Free Space on Your Phone

Does the same thing as tapping Clear Storage Space, as described in the previous section.

Change USB Connection Type

Lets you decide which USB mode your Fire phone uses when you connect it to a computer. In File Transfer mode, all your files are visible to the computer. In Photo Transfer mode, only photos and videos are displayed. For more details, see page 142.

Location Services

THIS LETS YOU TURN location services including GPS on and off and configure them for specific apps. With location services, your phone tracks your location so apps like Maps can use that information.

Configure Location Based Services for Your Applications

This turns location services on and off. If they're turned on, scroll down toward the bottom of the screen and you can then turn them on or off for each of the apps on your Fire.

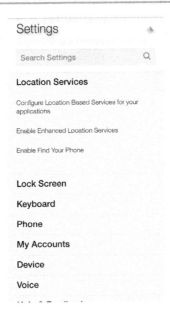

Settings

Search Settings

Location Services

Configure Location Based Services for your
applications

Enable Enhanced Location Services

Enable Find Your Phone

Lock Screen

Keyboard

Phone

My Accounts

Device

Voice

NOTE Whether you tap "Configure Location Based Services for your applications," "Enable Location Services," or "Enable Find Your Phone," you're sent to the same screen—the one that lets you customize all those settings.

Enable Enhanced Location Services

With Enhanced Location Services turned on, Amazon stores your location history and associates it with your Amazon account, which Amazon claims will enhance apps such as Maps and similar services. If you're worried about privacy, don't turn this on.

Enable Find Your Phone

Turn this on, and if you lose your Fire phone you can find it via the phone's location services. You can do more as well, including wiping data remotely and locking it remotely. If you lose your phone, head to your Amazon account on the Web and go to Your Account→"Manage Your Content and Devices." Click the Your Devices tab and select your Fire phone. Then click Device Actions, and you can remotely lock the device, locate it on a map, and sound a remote alarm.

Preston's 2nd Kindle

Preston's Kindle

Preston's Fire Phone

Preston's Fire Phone Edit

Device Actions ⬍   Email : preston_6493@kindle.com Edit

Deregister
Manage voice recordings
Remote Alarm
Find Your Phone
Remote Lock
Remote Factory Reset   s and Featured Recommendations
Manage Enhanced Location Services
Manage Firefly image and audio

Lock Screen

WANT TO CUSTOMIZE THE way the Lock screen looks and acts? Here's where to go.

Settings

Search Settings 🔍

Lock Screen

Select a lock screen scene

Set a password or PIN

Change the automatic lock time

Turn on notifications on the lock screen

Keyboard

Phone

My Accounts

Device

Voice

Help & Feedback

Select a Lock Screen Scene

Tap here, select a different scene for the Lock screen, and your will is done. If you're a fan of change, tap Rotate Scene at the top of the screen, and beneath that choose how often you'd like to have it rotate (never, every day, or every week). And if you want to have the scene be a photo you've taken with the Fire phone, tap Your Photo at the upper left and select the photo you want to use from your photo library (page 105).

Set a Password or PIN

Tap here and you can select a PIN or password for your Lock screen. That way, only someone who knows it can unlock your phone. (Traditionally, a PIN is numbers only, and a password is letters and other characters, but you can use any combination you wish.)

Change the Automatic Lock Time

You can choose how much time elapses before the screen locks when you're not using the phone. Tap here and select 30 seconds, 1 minute, 5 minutes, 10 minutes, or 30 minutes. You can also change whether pressing the power button locks the phone. (Out of the box, it does.)

Turn on Notifications on the Lock Screen

Normally, notifications aren't shown on the Lock screen. If you want them to show up, here's the place to make it happen. This screen also shows an overview of your current Scene, Screen Lock, and Automatically Lock options (tap to change any of them).

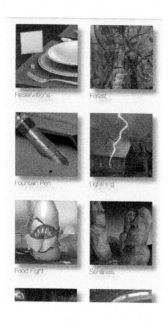

 # Keyboard

HOW MANY WAYS CAN you customize the way your keyboard works? Plenty, as you'll see in this section.

Change the Keyboard Language

You can select only between English and Spanish on the phone itself, but if you tap Download New Language you'll get a choice of dozens of options from Afrikaans to Vietnamese.

Keyboard

Change the keyboard language

Configure auto-correct and spell-checking

Manage advanced keyboard features

Edit your personal dictionary

Phone

My Accounts

Device

Change the date and time

Enable auto backups and backup now

Manage photo and video upload settings

Change your language

Install system updates

Configure Autocorrect and Spell Checking

Here you can turn spell checking and autocorrect on and off in most places that you type text—email and text messages, for example. But you have more options as well, like automatic capitalization for the first words of sentences, and being able to type on the keyboard by moving your fingers on it rather than having to tap—called Trace Typing.

Manage Advanced Keyboard Features

This is another way to bring up the "Configure auto-correct and spell-checking" screen. When you turn on Advanced Keyboard from this screen, when you hold your finger on a key on the keyboard, you'll get a list of additional characters you can use. For example, hold your finger on "h," and you can insert a < or : instead of the "h."

Edit Your Personal Dictionary

When you use words that the Fire phone's standard dictionary doesn't recognize, it flags them as misspelled. How annoying. Tap here, and on the screen that appears, tap the + sign. You can then add any words you want to the dictionary, which teaches the phone to recognize them as correctly spelled.

You can also delete words that you've accidentally added to the dictionary. To delete a word, tap the small pencil icon at upper right. Checkboxes appear next to all the words in your personal dictionary. Check any words you want to delete, and then tap the Trash icon at the bottom.

Phone

YOU REMEMBER TALKING ON a phone, don't you, rather than just using it for apps and the Web? Here's where to configure the Fire phone's many calling features. (For more detail on using these features, see Chapter 3.)

Configure Call Waiting

There's a lot more to call waiting than you might imagine, and this section lets you change it all. Here are the settings you can change:

Call waiting

This option lets you turn the call waiting feature (page 69) on and off.

Show my caller ID

Out of the box, your Fire uses your cellphone service provider's usual caller ID options—for example, if your provider has decided that your phone number appears on other people's phones when you call them. If you don't want to disclose your number, tap here and select Hide Number. If you want it to be shown, tap here and select Show Number.

Forward Incoming Calls

If you want incoming calls to go to another number, tap here. On the next screen, turn on Call Forwarding, and then type the number to which you want to forward your calls. To turn it back off, get back to the screen and turn it from On to Off.

Edit Reply-with-Text Messages

When you're getting a call and don't want to pick up, you can send the caller a text message instead. See page 66 for details.

My phone number

If you're having trouble remembering your own phone number (after all, you never actually call it), tap here for a reminder.

Voicemail number

Tap to see the number you need to call for voicemail. (See page 72 for details about how to set up voicemail.)

Voicemail password

You can change your voicemail password here.

View Your Phone Number

This section actually does more than just show you your phone number. You also get to see the software version of your Fire phone, its serial number, and more. For details, see the "Get info about your Fire" setting on page 326, because that brings you to the exact same screen as if you tap "View your phone number."

Contact Your Carrier

Tap here to see all the ways to get in touch with your carrier to obtain customer care, check your bill balance, get directory assistance, pay your bill, view your data usage, and see the number of minutes you've used in the current month. This contact information will vary, depending on your carrier (which, at this writing, could only be AT&T).

My Accounts

ACCOUNTS, ACCOUNTS, AND MORE accounts. Whether it's your email accounts, social media accounts, or Amazon accounts—among others—here's where you configure and customize them.

Deregister Your Phone

Click here, and you can unlink your Fire phone from your Amazon account. That way you—or someone else—can use a different Amazon account with it. After you deregister, all your Amazon-purchased content (books, movies, and so on) disappears from the phone.

This section also includes links for managing your Amazon account, social networks, email, contacts, and calendar.

Manage Email Accounts

This leads to a screen with links for managing not just your email accounts, but also your contact and calendar settings. From this screen you can add a new email account as well. (For more information about setting up email accounts, see page 193.)

You'll find plenty to customize for email and your contacts and calendar. With email, you get to change whether embedded images should be displayed; whether to automatically download attachments when you're connected via

WiFi; whether to include the text of messages when you reply to them; your email notification settings; and conversation settings, including whether to group them into discrete conversations.

For contacts, you can set whether to synchronize your contacts across all your Amazon devices; whether to sync your regular contacts with your Facebook contacts; and change the sort order of contact names and display by first name or last name.

As for calendar, settings include setting how much time ahead of appointments you should get reminders, the day you want your calendar week to start on, your time zone, whether to sync with Facebook events, and settings for your calendar notifications.

Connect Your Social Networks

This screen lets you link Facebook and Twitter to your Fire. For details, see page 222.

Manage Your Amazon account

This screen lets you change your Amazon account's name, email address, and password.

Manage Your Amazon Payment Method

Add, edit, or delete a method of paying on Amazon. Options include changing your "1-Click Payment" method and "Gift Cards & Promotional Codes" (redeem a code and view your gift card balance).

Manage Your Amazon Newsstand Subscriptions

Here's where you can manage—or cancel—your magazine and newspaper sub-scriptions (page 101).

Manage Your Send-to-Device Email Address

With your Amazon send-to-device email address, you (and contacts you approve) can send documents to your Fire just by sending them to this address. You can change this email address here. From this screen, essentially a web view of the one you'd get at Amazon.com, you can change the send-to-device email address (such as *joe.doe_36@kindle.com*) associated with any of your linked devices, not just the Fire phone. You can also enable or disable Personal Document Archiving and add or delete addresses from the Approved Personal Document E-mail list. (These are the email addresses you allow as *from* addresses, meaning that no one else can send documents to your device.)

Device

HERE'S THE PLACE TO go to manage everything from installing system updates, managing your SIM card, managing photo backup settings, and plenty more.

Change the Date and Time

In addition to changing the date and time, you can set your time zone, choose either standard or 24-hour format, and tell the Fire whether to automatically set its date and time from the cellular network or have you do it manually.

Enable Auto Backups

Tap here, and then turn on Device Backup to have backups done automatically. It backs up your overall settings, network settings, notes, web bookmarks, call history, and installed apps. To do a backup immediately, tap Back Up Now.

Manage Photo and Video Upload Settings

If you're on a plan that has a data limit, get to know these settings. They help you stay within your limit by automatically saving photos and videos to your Fire only when you're on a WiFi connection. That way, you'll have the content on your device and won't use any cellular data when you want to view the photos or videos. You can also decide whether to auto-save over a cellular network if

WiFi isn't available. (Out of the box this is turned off, so you won't use too much cellular data.)

The Uploads setting shows you any uploads you have queued up. Manage Storage tells you how many videos and photos you have on the phone, as well as how many you have in your Amazon cloud storage, called Amazon Cloud Drive. It also has a link to show you details about your current Amazon Cloud Drive subscription, so you'll know how much total storage you can use, and you can buy more cloud storage as well (Get More Space).

Change Your Language

Choose either English or Spanish.

Install System Updates

When you tap here, tap Check Now to see whether there are any new system updates. The screen also shows you your current Fire OS version, when it was last updated, and when it was last checked.

> **NOTE** Your Fire automatically checks for system updates on a regular schedule, so you only need to check for installed updates if you've heard one is available and want to see whether it's reached your Fire yet.

Factory Reset Your Fire

This scary option clears out every bit of information from your Fire phone and restores it to the state that it was in when it left the factory (page 298). Tap Reset on this screen, and then confirm that you want to do a factory reset. If you want to back up all your info, turn on the box next to Back Up Device Before Reset first.

Get info About Your Fire

Tap here, and the screen you come to has plenty of details, including things that make sense—phone number, device name, serial number, and software version— and things you may not understand—IMEI number and Wi-Fi MAC address. This is a good screen to know if you need to call tech support, because they may ask for this information.

Configure Your Emergency Alerts

Your Fire phone can automatically receive many kinds of emergency alerts, including threats to life and property and child abduction emergency bulletins called Amber Alerts. Here's where to turn them on and off and set whether your phone should vibrate for each emergency alert sent to you.

View Your Emergency Alerts

Tap here to see all the alerts that have been issued to you.

Manage Your SIM Card PIN

Tap here to require a PIN to be entered for your SIM card, and to change the PIN. This PIN provides an additional level of security for your Fire phone, because it means that someone will have to enter a PIN for the SIM card, as well as a PIN or password to unlock the screen if you've set it up that way (page 317). Also, by creating a PIN for the SIM card, you ensure that the card can't be taken and used on another device.

Manage Enterprise Security Features

This screen isn't really for you. Instead, your company's IT gods will use it for things like storing credential storage VPN information and more.

Manage Accessibility

If you have accessibility issues, here's the place to go. You can enable a screen reader if you have vision problems, use a screen magnifier, turn on spoken Caller ID, and more.

View Legal and Compliance Info

In the mood for a snooze? Then tap here and read scintillating text like terms of use.

 # Voice

THERE'S NOT MUCH HERE—JUST two screens of settings.

Configure Voice Settings

From here, turn voice input on and off, select your language, and turn hands-free mode on and off. You can also disable voice input and change speaking speed from this screen. (For details about using voice controls, see page 82.)

Change Text to Speech (TTS) Language

Your Fire can read the text on your phone and turn it into speech. Out of the box, it uses English. But if you want to use another language, tap here and then tap Download New Voice, and you can choose from many other languages. There are lots of them in addition to the usual. Welsh, Danish, or Turkish anyone?

 # Help & Feedback

Get Help From Mayday

This option launches the Mayday app that connects you to live video tech support. For details about how to use it, see page 36.

Browse Online Help

This item launches a user guide for the Fire phone. You can either browse or search through its contents for the help you're looking for.

Contact Amazon Technical Support

Tap this and you'll get two choices: Customer Service and Feedback. Tap Customer Service, and you can email or call Amazon. Tap Feedback, and you're led to a screen that lets you tell Amazon which feature of the Fire phone you're having problems with.

Provide Feedback

This leads you to the same screen you get to if you tap "Contact Amazon technical support"→Feedback.

Searching Settings and Recent Settings

DRILLING DOWN INTO THE depths of Settings can be a time-consuming bore. Luckily, there are several quicker ways to get to the settings you want. One is to do a search. Scroll to the top of the screen, and in the Search Settings text box, type the kind of setting you're looking for. The Fire phone displays matching results. Tap any setting to head to it.

You can also swipe or tilt to open the right panel and see all the settings you've most recently used. Tap any to go to it.

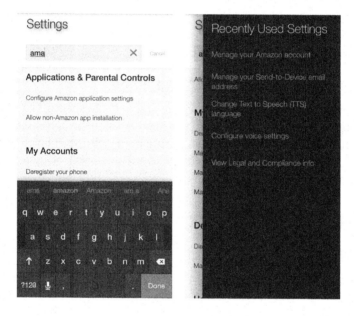

Index

onscreen controls (Camera app), 117–121

Oovoo app (videochat), 77

organizing information with Evernote, 273–274

Orientation tool (photos), 110

Overdrive software, 101

P

pages, turning (Kindle app), 90–92

pairing Bluetooth earpiece with phone, 75–76

panning photos, 109

Panorama mode (Camera app), 119

parental control settings, 311–312

passwords

    e-mail accounts, 198

    Lock Screen, 317

    mobile hotspot, 162

    Remember passwords setting, 190

    voicemail, 321

PCs, transferring files using, 142–144

Peek feature

    for book information, 98

    for magazines, 101

    in Maps app, 251

Peek gesture, 9, 23–25, 106

permissions (apps), 260

personal dictionary setting, 319

Personal Document Archiving, 323

Phone app

    answering calls, 65–67

    Bluetooth technology, 74–76

    caller ID, 71–72

    call forwarding, 70–71

    call settings, 320–322

    call waiting, 69–70

    conference calls, 67–68

    overview, 59–60

    ringtones, changing, 73

    text messages. See text messages

    VIP list, 64–65

phone numbers

    old vs. new, 288

    remembering personal, 321

    using Firefly with, 55–57

photos

    adding to contacts, 210

    filtering/sharing via Instagram, 275–276

    lenticular, 119

    panning, 109

    Photo Transfer setting (Macs), 142

    selecting for Lock Screen, 317

    sending in text messages, 81–83

    sending via tweets, 238

    shooting still, 116–117

    uploading to Facebook, 228–230

    Uploads setting, 324–325

    zooming, 108–110

Photos app

    basics, 105–108

    left/right panels in, 112–113

    multiple photos, 111

    Photo Editor, 108

    viewing/manipulating photos, 108–111

Pictures folder, 144

Pinch and Spread gesture, 21

Amazon Fire Phone

THE MISSING CD

There's no CD with this book; you just saved $5.00.

Instead, every single Web address, practice file, and piece of downloadable software mentioned in this book is available at *missingmanuals.com* (click the Missing CD icon). There you'll find a tidy list of links, organized by chapter.

CPSIA information can be obtained at www.ICGtesting.com
Printed in the USA
BVOW09s0425111214

378927BV00005B/5/P